The Three Big Bangs

THE THREE BIG BANGS

Comet Crashes, Exploding Stars, and the Creation of the Universe

PHILIP M. DAUBER
RICHARD A. MULLER

HELIX BOOKS

Addison-Wesley Publishing Company, Inc.
Reading, Massachusetts Menlo Park, California New York
Don Mills, Ontario Harlow, England Amsterdam Bonn
Sydney Singapore Tokyo Madrid San Juan
Paris Seoul Milan Mexico City Taipei

Library of Congress Cataloging-in-Publication Data
Dauber, Philip M.
The three big bangs : comet crashes, exploding stars, and the creation of the universe / Philip M. Dauber and Richard A. Muller.
p. cm. — (Helix books)
Includes index.
ISBN 0-201-40752-3 (hardcover)
ISBN 0-201-15495-1 (paperback)
1. Big bang theory. 2. Catastrophes (Geology) 3. Supernovae.
I. Muller, R. (Richard) II. Title.
QB991.B54D38 1996
523.1'8—dc20 95-24451
CIP

Cover design by Lynne Reed
Text design by Diane Levy
Set in 11-point Minion by Compset, Inc.

1 2 3 4 5 6 7 8 9 10—DOH—0100999897
First printing, November 1995
First paperback printing, March 1997

Contents

III: Creation of the Universe

Preface

This account of the physical origins of life on Earth focuses on three crucial and violent events. Almost everyone has heard of the first one, but few understand it: the creation of the universe, as scientists today describe it in terms of Big Bang theory. The second and lesser-known big bang is the supernova, the cataclysmic stellar explosion in which the chemical elements that compose our world and our bodies were formed. The third big bang is the impact of a comet or asteroid crashing onto the Earth, extinguishing some species and allowing others to thrive. This big bang happened some 65 million years ago, and it wiped out the dinosaurs and triggered the rapid expansion of mammalian species that led to human beings. Probably such bangs have happened many times during our prehistory. If so, the violent extraterrestrial impact must be counted as one of the main driving forces of biological evolution, perhaps as important as competition among species. In July 1994, a comet that crashed into Jupiter with spectacular results reminded us of the awesome power of planetary impacts.

To make the book as accessible as possible to the nontechnical reader, we have chosen to tell our tale in reverse chronological order, beginning with the impacts on Jupiter and Earth and ending with Big Bang cosmology itself. The first part of the book deals with life-and-death dramas experienced by living creatures, while the later parts are concerned with violent happenings in the cores of collapsing stars or in the early universe before stars were even able to form. After the overview of the book in Chapter 1, Chapters 2 through 10 explain the evidence for catastrophic impacts and their implications for the evolution of life. Chapters 11 through 15 mainly cover supernova explosions, while Chapters 16 through 21 summarize Big Bang cosmology, emphasizing its roots in Einstein's relativity theory and the observational evidence for it. Chapter 22 shows how knowledge about supernovae may help scientists solve some of the most enigmatic puzzles about the universe, after which Chapter 23 reprises the main themes of the book and looks ahead to future discoveries.

Today a grasp of the basic facts of biological evolution is considered essential for a well-educated person. It is no less essential to understand the main stages in the *physical* evolution of matter and energy. Bearing the general reader in mind, we have structured our story to communicate a sense of deepening mystery, as if it were a suspense novel. But we also intend that the book be used as a supplementary text in physics and astronomy classes. To keep it manageable both in and out of classroom situations, we have kept the chapters relatively short and organized the material into easily digestible chunks.

The Three Big Bangs is not intended to be an up-to-the-minute account of all the latest ideas in cosmology, or in impact and supernova research. Startling observations by a single research group are sometimes refuted when the astronomical community attempts to confirm them. Speculative new theories in these fast-moving fields have even lower survival rates. While controversies rage, we have chosen to focus less on them than on the marvels that we do know about the three big bangs.

The Three Big Bangs

CHAPTER ONE

The Three Big Bangs

In this book we are going to ask you to imagine a series of events so violent, they dwarf the most vicious crimes that humans have committed against humans and the most horrible natural catastrophes that have taken place on Earth. Even the puniest of the three big bangs, the crashes of comets onto the Earth many millions of years ago, released a destructive energy greater than would be unleashed by simultaneous detonation of every nuclear weapon ever made. In fact, its energy would surpass such a nuclear holocaust by more than ten thousand times.

The challenge to the reader of this book is to come to grips with these horrendous events. For only by understanding them can we really understand our origins.

In school we learn of biological evolution: how species competed with other species, often violently, until the weaker ones were eliminated. But in just the last decade, new discoveries have challenged this understanding of biological evolution. More remarkably, scientists have recently begun to understand our *physical* evolution, to the point that we can now speak in a meaningful way about the origins not only of our countries and our cells but of our very atoms. Perhaps most surprising, we are even beginning to understand the origins of the universe itself, which according to current theory involves the creation not of matter alone, but of space itself and maybe even of time.

The physical creation of the world we know is dominated by violence—violence on a scale so far beyond human dimensions that some consider it impossible to imagine. In recent years we have begun to recognize that the violence of nature is key to answering the otherwise impossibly difficult question: How did we get here?

Fascinating to both children and adults, this question lies at the core of myths and religions in primitive and advanced civilizations alike. Perhaps your parents told you, "We were created by God." Even if this answer were literally true, it would not be entirely satisfactory, because it doesn't explain *how* God created us.

Scientists have often underestimated the role of large-scale sudden violence in nature, for the simple reason that such events are so rare, we have little experience with them. Their very infrequency means that they are not part of our imagination. We are accustomed to thinking of evolution, for example, as gradual. The evolutionary changes that Darwin saw were slow, like the colors of surviving butterfly species that are altered to match their changing environments. But more recently, two outstanding paleontologists, great fans of Charles Darwin, have argued that the theory of evolution may require a major overhaul. Stephen Jay Gould and David Raup, renowned experts on early life and evolution, argue that the great changes in species may have been driven less by day-to-day competition than by extremely rare and violent events.

Our normal vocabulary is inadequate to name such destructive events. For lack of a better phrase, we use the term originally applied to a particular cosmological theory: *big bang.* (Fred Hoyle originally coined the phrase to make fun of the new theory of his friend George Gamow.) For the results of such events, however, we do have a name: We call them *mass extinctions,* for most life-forms on earth are completely destroyed by such events.

This book is about three big bangs. The first—closest to the human scale—is the big bang that took place 65 million years ago. One day without warning, a comet, or possibly an asteroid, smashed into the Earth, forever changing life on our planet. The impact blasted out a giant crater in the present-day Yucatán, in Mexico. The atmosphere, the oceans, forests, and jungles were totally disrupted in ways that scientists are now busy unscrambling. Dinosaurs and most other life-forms vanished, including most existing mammals. But some mammals, our ancestors, survived and went on to flourish. There have been many such biological catastrophes, but the one best understood, thanks to a series of remarkable discoveries over the last fifteen years, is this Cretaceous-Tertiary catastrophe.

To astrophysicists, the impact of a comet on Earth is a puny event—tiny compared with that of a star that erupted 5 billion years ago, a cataclysm even more important to our evolution in the sense of physical rather than biological evolution. Where biologists ask "How did life arise and become what it is today?" physicists ask how the matter we are made of was created, how it changed over the aeons, and what forms it takes now.

When the first stars formed, you could not have found the atoms that now make up your body, but you could have discovered their precursors buried deep inside the stars. Many of these atoms were totally unrecognizable. The iron of your present-day blood, for example, was not yet iron; mostly it existed as hydrogen and helium. Unformed also were the carbon, nitrogen, and oxygen in your organic molecules. In the next few billion years, the nuclear forge of a star would cook hydrogen and helium, through thermonuclear fusion, to create new atoms. But these atoms were still buried in the star. In the second big bang, these atoms were created and ejected into space.

This big bang preceded the extinction of the dinosaurs by about 5–10 billion years. With little warning, the star erupted, blasting the new atoms over a region of space hundreds of light-years in size. It was a *supernova,* and without it no life would have appeared in our part of space, for none of the elements that make life possible would have been present. From the ashes of this supernova a second star eventually coalesced, a star that someday would be named the Sun by the two-legged creatures whose bodies were formed from the atoms forged inside the supernova and who now inhabit the small water-cloaked planet that formed near the Sun.

Our third big bang is *the* Big Bang, the one you read about in the newspapers and science magazines and that long preceded the other two. It was the ultimate explosion that involved all the energy of the universe, the explosion that can never be exceeded, the violent event that dwarfs all other violent events. Although the great scientist George Gamow originally conceived this Big Bang as the event in which all the elements of the universe were created, we now know that most of them, except for hydrogen and helium, were created much later, inside stars.

The story of *the* Big Bang has evolved in ways that few people could have foreseen more than forty-five years ago, when the idea was first set forth. We now understand the Big Bang as the event in which hydrogen and helium were created from more primitive particles—and something much more fundamental and mysterious as well. Here is a concept even more jarring to our minds than the creation of all matter: The mind-wrenching idea that makes the Big Bang so fascinating is that it represented not only the creation of matter within space but the creation of space itself. And since the Big Bang was the creation of space, then by our current understanding of relativity theory it may also have been the creation of time.

These great catastrophes played a role in our physical and biological evolution that is only now being recognized. Scientists overlooked them for a long time, we think, because catastrophes are rare events, very far from our usual experience. They learned to explain continuous change with the mathematics of Newton and his successors. But now in the late twentieth century, with all other explanations exhausted, scientists finally stretch their minds to calculate the unimaginable. Catastrophe is at the forefront of science today because it is a mystery left over from the triumphs of science through the midtwentieth century. (Chaos is another such mystery.) Much more difficult to understand than Newtonian mathematics, catastrophe has been left for us to unravel now.

The authors of this book have had the wonderful privilege of being able to research each of the three big bangs. ("Our research has been one catastrophe after another," we sometimes joke.) Although the three big bangs may not seem to have much to do with each other, they do. The strongest link tying them together is their deep shared relevance to the roots of all life on Earth. When we study the comet impact, the supernova, and the Big Bang, we are studying our common history, our truly ancient history. What has drawn us to all of these events is a deep desire to find out *where we came from.*

I

Comet Crashes

CHAPTER TWO

Jupiter Crash

Never before had astronomers witnessed such catastrophic violence so close to Earth. Never before had so many telescopes been aimed at a single target. Not since the invention of the telescope more than three hundred years ago, and its first use by Galileo had there been such revelation from the sky. Beginning on July 16, 1994, twenty-one comet fragments slammed into the planet Jupiter at a speed of nearly 60 kilometers per second—about sixty times faster than a rifle bullet. So spectacular were the results of the impact that even amateur astronomers with small backyard telescopes could witness it with their own eyes. Large telescopes revealed, in incredible detail, a series of big bangs so great that had they befallen the Earth, civilization as we know it would have been wiped out, and possibly all human life.

The largest fragment, estimated to be 3 to 4 kilometers in diameter, exploded in a brilliant fireball nearly the size of the Earth. Its impact energy was the equivalent of about *6 trillion* tons of TNT, thousands of times more than would be released by the simultaneous detonation of all existing nuclear weapons. (In scientific notation, 6 trillion is 6×10^{12} or, on your pocket calculator, 6E12. In either case it's a 6 followed by twelve zeros.) The fireball swirled into view only minutes after the fragment's impact, a bright ball of mostly infrared radiation. Minutes later, the fireball faded, leaving a dark spot surrounded by thin concentric rings, possibly due to expanding sonic booms. Like some twenty other new scars on Jupiter's atmosphere, this Fragment G site remained visible for months. Formed by sulfurous dust blown away from the point of impact, this largest impact spot spanned an area more than two Earth diameters across.

Jupiter is a remote world, very different from our rocky, water-covered Earth. As seen from Earth, it is the third-brightest object in the night sky—only the Moon and Venus appear brighter. Most of the giant planet is probably liquid hydrogen, wrapped in thick clouds of gaseous hydrogen, helium, methane, ethane, carbon monoxide, and hydrogen cyanide. The very top layer is rich in crystals of frozen ammonia. Deeper down there is water, both in ice-crystal and in liquid form. Astronomers now have evidence for the presence of sulfur-containing compounds such as ammonium hydrosulfide as well.

When each comet fragment smashed into Jupiter's upper atmosphere, it shock-heated the planet's atmospheric gases to a temperature of many thousands of degrees, hot enough that the gases glowed brightly. The Galileo spacecraft, 150 million miles away, observed these initial flashes directly. After the fragment exploded in a giant fireball, observers on Earth usually had to wait a few minutes for Jupiter's rapid spin of ten hours per revolution to bring the fireball into view. But around the world they saw, over Jupiter's horizon, tall plumes of glowing gas that had been created with some of the fireballs. As the plume material dropped back to Jupiter's atmosphere, it heated gas molecules again, producing bright spots at infrared wavelengths that amateur and professional astronomers could see but that at visible wavelengths looked dark. At the impact spots, scientists detected the first hydrogen sulfide gas (the noxious substance that gives rotten eggs their unpleasant smell) observed on Jupiter, as well as other sulfur-containing molecules.

Scientists aboard NASA's Kuiper Airborne Observatory also detected water from the impacts. For a typical impact, the amount of water was about what would be expected from a ball of ice 400 meters in diameter. But had the water they observed come from a comet fragment or from Jupiter's own atmosphere? The scientists still aren't sure.

After a week of comet crashes, Jupiter's battered southern hemisphere was sprinkled with more than a dozen spots, each marking an impact site.

How do such planetary disasters occur? How often do comets or other objects from space hit planets? Is Earth as vulnerable to collision as Jupiter? What would happen to us in the event of a cosmic crash? Coincidentally perhaps, research of the past fifteen years has enabled scientists to answer such questions with increasing authority. In the light of what is now understood, the amazing events of July 1994 are a warning to us that our planet is not as safe as we once thought.

Known as comet Shoemaker-Levy 9, the train of objects that smashed into Jupiter were discovered in March 1993. Amateur astronomer David Levy and the husband-wife team Carolyn and Eugene Shoemaker had been searching for comets and other "near-Earth objects" for many years. They had stayed up all night photographing the same segments of sky, year after year, waiting for some previously unknown chunk of ice, rock, or other matter to make a dramatic appearance in the inner solar system. Comet hunting, like much of modern science, is a highly competitive game. Levy and the Shoemakers were good at it; some say they were the best. Between them they have discovered many dozens of these icy chunks with glowing heads and long tails.

But on the night of March 24 they were lucky—very lucky. They were using one of the smaller wide-field telescopes at Palomar Observatory in southern California. Visibility was poor, and they were running low on good photographic plates. In fact, the sky was so cloudy, they debated whether they should continue observing at all. But then Levy found some damaged film that had previously been exposed accidentally. He decided to use it up—if they didn't find anything, it wouldn't be much of a loss. Were it not for his optimism and thrift, scientists might have been taken by surprise by the Jupiter impacts of July 1994 and possibly never understood them. The team took a few pictures, then quit for the night.

Next day, the Levy-Shoemaker team scanned the pictures. Despite some blurring, they found an object quite unlike any they had ever seen before, not far from Jupiter. Unusually long, narrow, and fragile-looking, it had a tail like a comet. But was it really a comet? Unable to look again at their strange discovery because of continued cloudy skies, the team called on Jim Scotti, who uses the 0.9-meter (36-inch) Spacewatch telescope at Kitt Peak National Observatory in Arizona, to search for asteroids on near-collision courses with the Earth. With this powerful instrument, Scotti was quickly able to image the new object, using not photographic film but a digital "camera." Yes, it was a comet, but one apparently made of many fragments strung out for hundreds of thousands of kilometers.

When astronomers aimed very large telescopes at the latest Levy-Shoemaker discovery, they counted twenty-one fragments spread out in an almost perfectly straight line. More startling, they discovered that this "string-of-pearls" comet was in orbit not around the Sun, as are most comets, but around Jupiter itself. Evidently the giant planet had, possibly within the previous ten years, "captured" the comet with its powerful gravitational field. Calculations at the Jet Propulsion Laboratory in Pasadena, California, showed that the comet's orbit has taken it as distant as 31 million miles from Jupiter, and as close as 16,000 miles, where tidal forces arising from the giant planet's gravity had torn the comet into pieces on July 7, 1992. Now, the calculations showed, the comet was doomed to smash into its captor in July 1994.

What, scientists wondered, would happen when comet Shoemaker-Levy 9 hit? What, if anything, would we see from Earth? Remembering the debacle of comet Kohoutek in 1973, when what was predicted as the show of the century turned out to be a giant fizzle, astronomers were relatively cautious about issuing forecasts. Kohoutek had been unusually bright when still at a great distance from Earth, but when it got closer, it was barely visible, even with large telescopes. For Shoemaker-Levy 9, predictions for the impending crash ranged from "nothing will be visible" to "monumental fireballs," "giant mushroom clouds," and "Jupiter will light up like a Christmas tree from cometary dust." Some scientists doubted that any explosions or lingering effects would be seen at all, other than by the most powerful professional telescopes. Hardly anyone predicted confidently that every backyard astronomer in the world would be able to see the impacts with ease. After all, Jupiter would be 770 million kilometers away during the week of the impacts. Besides, there was evidence the comet fragments were breaking up; the orbit calculations might turn out to be wrong.

So it was that astronomers both amateur and professional were filled with joy when the long-anticipated fireballs and dust clouds actually materialized. One of the authors happened to be in Boston during the week of July 16. Traditionally, amateur astronomers set up their telescopes once a week on the top level of the parking garage at the Boston Museum of Science and proudly offer the public a chance to view the heavens. On July 18, huge crowds showed up for that amateur night. To get a peek through any of the dozens of fine telescopes, you had to wait in a long line. Clouds obscured the sky once again, and city lights and haze made

Jupiter hard to see at all, but the excitement of the crowd was palpable. Through every one of the telescopes, impact spots could be seen clearly; if you could see Jupiter at all, you could see at least one spot.

The scars lasted longer than most astronomers expected. Jupiter's rapid rotation and 300-mile-an-hour winds, it was guessed, might have quickly torn the spots apart and dissipated their contents. Weeks later, however, some scars endured, albeit partially smeared and twisted. Within Jupiter's stratosphere, as in the upper atmosphere of our own planet, there is little vertical movement. (On Earth, the anvil heads of thunderstorm clouds last for hours, not months, and volcanic dust, spewed into the atmosphere by a violent explosion, can darken sunsets for years.) Months after the crash on Jupiter, the spots had merged into long streaks that completely encircled the planet.

From the size of the spots, astrophysicists could calculate that the energies involved in the impacts were the equivalent of thousands of megatons of TNT. This confirmed the scientists' calculations of the size and mass of the comet pieces; comet cores are indeed several kilometers or more in diameter. (This fact, as we shall see, supports the theory that impacts of extraterrestrial objects were responsible for the mass extinction of life on Earth, including all the dinosaurs, 65 million years ago.)

The message of comet Shoemaker-Levy 9 is all too obvious, as numerous scientists and journalists pointed out after July 1994. If a visitor from the depths of the solar system, such as a comet or asteroid, could cause such devastation on mighty Jupiter, we on Earth are even more vulnerable. It is much less probable that Earth could *capture* a comet with its gravity, as did 318-times-more-massive Jupiter, but capture is not necessary for impact.

Although the speed, energy, and momentum of comets are impressive and the effects of their impact spectacular, they are much smaller in size than planets. At eleven times the diameter of Earth, Jupiter is 143,000 kilometers across, or about fifty thousand times the span of the largest comet fragment. Jupiter's mass is more than a hundred million million (one hundred trillion, or 10^{14}) times greater than that of the comet. Fears that Jupiter (or even a much smaller planet like Earth) could be knocked far out of orbit by such a comet are completely baseless.

Yet Earth has already been struck by comets and asteroids that have profoundly changed the course of our natural history. There are billions of comets in the solar system, most of them too far away to be seen by even our largest telescopes, but any one of them is a potential killer. Thousands of asteroids have crossed Earth's orbit in space, potentially on a collision course with us. Evidence collected in the last few decades makes it impossible to maintain that the hundreds of huge impact craters that mar the surfaces of the Moon, Venus, and other planets are all of volcanic origin, as some geologists once insisted. More than a hundred large impact craters, long hidden by erosion or the ocean, have now been discovered on Earth itself. In 1908 a giant explosion rocked a remote region of Siberia, knocking down trees for miles and releasing the energy equivalent of a 10-megaton thermonuclear bomb. The only known explanation for this disaster is the impact of an extraterrestrial object. The evidence is incontrovertible: We live in a giant shooting gallery, and we are a target.

As you go about your business tomorrow, think about this: Only long odds protect you from an asteroid or comet strike. Surprisingly, compared with other daily hazards we all face, the odds are not so long.

CHAPTER THREE

Target Earth

It was a day much like any other day 65 million years ago, except for one curious difference: There was a small bright spot in the sky that kept getting bigger and brighter. The spot was a 6-mile-wide asteroid or comet on a collision course with Earth.

Four hours before the impact, the approaching killer was about as far away as the moon and about as bright as Venus at twilight.

Ten minutes before impact, it was within one Earth diameter. Whether any of the doomed creatures on Earth took notice of it, we cannot know. Had humans been around, they could have observed the shape of the object by then; probably it was quite irregular. Perhaps they could have seen it tumble. Had it been a comet, the glowing head would have loomed large, and the multicolored streaked tail pointing away fom the Sun would have been spectacular.

Ten seconds before impact, the invader blazed with a fiery glow as it entered the upper atmosphere. A yellow cylindrical contrail formed behind it, expanding at the speed of sound. Some of the alien material vaporized; some was knocked off as dust. But the main mass of the asteroid or comet crashed into the ocean, penetrated to the sea floor in less than a second, and blasted its way through mud and ooze into the Earth's crust.

The Cretaceous era of Earth history, the age of the dinosaurs, was over. Amidst unimaginable violence, the Tertiary period had begun, later to be dominated by our ancestors, the early mammals.

Within a few seconds of the impact, energy equivalent to millions of thermonuclear bombs was released, much of it as heat. Temperatures within a few hundred meters of the impact soared to over 1 million degrees Celsius. Mud, water, and even some rock vaporized; more rock melted. A huge fireball exploded out of the sea in ghostly slow motion, because of its enormous size. In truth it carried destruction with it at faster than the speed of sound.

Expanding through the Earth at more than 4 kilometers per second, the shock wave dug out a monstrous crater nearly 200 kilometers wide. Earthquake tremors roared outward from the epicenter. Even those dinosaurs and other animals that took it as a warning could do little to protect themselves.

Rebounding chunks of asteroid and debris hurtled upward and outward. The mass of dust alone totalled 100 trillion tons—the equivalent of a billion large ships. Countless pieces shot far out into space as glowing meteors. These missiles cooled temporarily, then flared up again as they rained back down to Earth.

Forests and jungles within a thousand miles or more burst into flame. If any burning trees were left standing, a blast wave of pressurized air knocked them down. As secondary fragments bombarded the Earth, distant forests and grasslands also burned. Intense heat from the reentering meteors baked many animals alive.

In the oceans, the crash spawned huge tsunami waves that sped outward across the sea at hundreds of kilometers per hour. A thousand miles away from the impact, towering walls of water taller than any building standing today roared over beaches and crashed over hills, smashing everything in their path. Coastal plains that had nurtured abundant life uneventfully for millions of years were completely inundated.

Hundreds of square kilometers of ocean endured severe heating. Near ground zero, the sea flowed into a superheated cavity and boiled. Above the overheated ocean, a storm formed, a runaway hurricane greater than anything humans have ever witnessed. Driven by the abnormally large temperature difference between warm water and frigid stratosphere, winds built quickly to more than 800 kilometers per hour. Updrafts loaded with water vapor thrust as high as 50 kilometers, disrupting the upper atmosphere. So powerful was the storm, which scientists have dubbed a *hypercane,* that its top winds may have reached supersonic velocity. It lasted for days, while the ocean surface slowly cooled. Many cubic *miles* of water may have been lofted into the stratosphere, enough to drastically change world climate.

More cubic miles of dust fell back to the stratosphere and, carried aloft by winds, spread throughout the planet. Everywhere, day gave way to inky night. The Sun and Moon would not be seen clearly again for months. Not a single star was visible. Temperatures plunged worldwide, in some places from oppressive heat to subfreezing cold.

Photosynthesis by ocean plankton ceased. Most marine life, dependent on plankton at the base of the food chain, was doomed. Months after the impact, when the dust finally settled, a dense haze of sulfuric acid droplets may have remained, formed from a hundred billion tons of sulfur compounds spewed into the air when the asteroid hit. (Much rock is high in sulfur content.) The fiery, explosive cauldron of the impact released enormous quantities of sulfur dioxide. Reacting with billions of tons of water vaporized by the fireball, this noxious gas formed a yellowish acid mist that spread throughout the stratosphere. Sulfuric acid clouds would continue to filter out sunlight for decades. Most land plants that survived the firestorms would die in the cold and darkness; with them, many animals would perish. Those that survived would have to contend with yet another horror: acid rain.

The intense heat of the fireball fused vast amounts of atmospheric oxygen and nitrogen into oxides of nitrogen. Today oxides of nitrogen from automobile exhaust are a principal cause of smog. Reacting with water in the air, these compounds form nitric acid, which along with sulfuric acid is one of the most corrosive substances known to chemists. In the aftermath of the asteroid impact, acid rain fell everywhere on Earth, in concentrations far greater than those that damage forests today. The acid rain may have been sufficient to kill much of the remaining plant life. Ocean acidity rose to the point where many forms of plankton could not survive; the forms that did are those more resistant to high acidity.

One of the most common rocks on Earth is limestone, which is mostly calcium carbonate. In the colossal explosion of the asteroid or comet impact, much of the carbonate decomposed, and carbon dioxide spewed into the atmosphere, greatly increasing the amount present at that time. Carbon dioxide and water vapor in the atmosphere acted to trap the

Sun's heat, producing the so-called greenhouse effect. When the cooling dust, water, and acid clouds finally settled out of the atmosphere, the excess carbon dioxide and water vapor remained. Earth's climate may then have rebounded from extreme cold to extreme heat. Not until the surviving green plants recovered and restored the atmospheric balance could the climate have returned to normal. (Photosynthesis consumes carbon dioxide.) This process might have taken thousands of years.

So disrupted was Earth's atmosphere that most of its ozone layer was destroyed. Normally, atmospheric ozone performs the vital function of screening out the Sun's ultraviolet (UV) radiation. But as Sun lovers today have become aware in recent years, even the tiny fraction of UV that penetrates the ozone layer can cause skin cancer and eye damage. Without an atmospheric ozone shield to protect them, many species were vulnerable to extinction.

This frightening picture of ancient catastrophe is not mere speculation but is supported by the fossil record from 65 million years ago. In one of the greatest mass extinctions of prehistory, over two-thirds of all existing plant and animal species disappeared. Not a single land animal species weighing more than a medium-sized dog survived the crisis. Every single species of dinosaur vanished, except for birds, which some scientists think are descended from the dinosaurs. Many existing mammalian species died off too. The killing was even more complete in the oceans, where it included most microscopic forms of life. Paleontologists have even found indications of the very rapid climate zig-zags that a large impact could cause.

How certain are we that this asteroid or comet crash really happened? Have extraterrestrial bodies big enough to cause mass extinction actually crashed into the Earth?

A few scientists had recognized the hazard of asteroid impacts ahead of their time. As early as 1941, Fletcher Watson estimated their frequency based on the discovery of the first Earth-approaching asteroids. Ralph Baldwin warned in his 1949 book *The Face of the Moon* that the explosion that had formed the lunar crater Tycho "would, anywhere on Earth, be a horrifying thing, almost inconceivable in its monstrosity." During the 1970s the well-known Canadian paleontologist Digby Mclaren suggested that a giant meteorite had caused a mass extinction 365 million years ago. Earlier, Irish comet expert E. J. Öpik had published papers speculating that comets could annihilate life in large regions and possibly destroy species. And in 1973, Nobel-laureate chemist Harold Urey published a paper claiming that comet impacts had caused lesser extinctions during the past 50 million years. He speculated that a comet had been responsible for the extinction of the dinosaurs and even suggested that tektites dating from the end of the Cretaceous would eventually be unearthed. Despite the eminence of these authors, no one paid much attention to their warnings and proposals. What was lacking was that scientific essential: evidence.

Some great discoveries in science are accidents, like Sir Alexander Fleming's discovery of penicillin. Others occur as a result of a patient search with conventional technology, like the detection of comet Shoemaker-Levy 9. Still others reward the builders of new, bigger, or more sensitive scientific instruments such as the Hubble space telescope. But other discoveries, sometimes those hardest won, come after a long struggle of puzzle solving, requiring both luck and extraordinary skill. So it was with the discovery that something huge from space had smashed into the Earth around the time the dinosaurs became extinct.

The scientists most closely associated with that discovery are Luis and Walter Alvarez. In 1977, Walter Alvarez was visiting Berkeley for a year. A geologist at Columbia University's Lamont-Doherty Geological Observatory, he was thinking about accepting an assistant professorship at the University of California, albeit at lower pay. It was not a decision to be made lightly. Favoring the move was the fact that Walter's own father, Luis, was a physics professor at Berkeley. Luis had won the Nobel prize in physics in 1968. Walter had never worked professionally with his famous father, but the idea of doing so intrigued him.

To physicists at Berkeley, Luis Alvarez was an awesome figure, both superstar physicist and stern taskmaster. Yet even young graduate students and postdocs called him Luie. Both authors of this book were Luie's protégés and have been profoundly affected by our contact with him.

Luis Alvarez won the Nobel prize for his group's discoveries of elementary particles with the bubble chamber, findings that led to the current "standard model" of subatomic matter. But he made many other important discoveries as well. He discovered the basic radioactive phenomenon of electron capture. He found the radioactivity of tritium, the rarest isotope of hydrogen, and also the magnetism of the neutron. He showed that most cosmic rays are protons.

In addition, Luie was also an accomplished inventor who had been elected to the Inventors Hall of Fame. He invented the trigger for the atomic bomb and the first method of landing airplanes relying strictly on instruments. He also used cosmic radiation to "X-ray" the pyramids of Egypt and analyzed the Zapruder film of President Kennedy's assassination so incisively that CBS television based a series of TV specials on his conclusions.

Walter Alvarez finally did decide to go to Berkeley, and when he arrived, he brought his father a scientific gift. Within the gift, Walter suspected, was the key to the extinction of the dinosaurs. It was a small section of sedimentary rock that he had cut from an outcrop near Gubbio, Italy. To keep the rock from crumbling, Walter had encased it in plastic. Walter had Luie look, with a magnifying glass, at a wide variety of small fossils, called forams, in the light-colored limestone layer at the bottom. Above the limestone was a dark layer of clay, and above that another layer of limestone. The upper limestone layer contained virtually no forams. Each layer had been formed by tiny particles settling in the ocean. Evidently, in the time interval between the deposition of the bottom layer and that of the top layer, all the many types of forams had been wiped out by some unknown catastrophe. And, Walter pointed out, so had the dinosaurs.

The pattern of layers that Walter showed his father is found in deposits all over the world. Dinosaur fossils, with their large bones, appear in abundance below the thin dark clay layer. Above that layer, there are virtually none. (There are no known complete skeletons, but miscellaneous bones have been washed into the younger layer following Earth movements.) Whatever had wiped out the forams, Walter speculated, had done in the dinosaurs too.

Luie had heard about this great mystery of geology and paleontology, but now, with the evidence in his own hand, he was fascinated. What, he wondered, had made that layer of clay? Had it been deposited in one year, a hundred years, or a hundred thousand?

Several years before Luie got interested in the dinosaur problem, one of the authors, Richard Muller, had worked with Walter to solve it by measuring the number of atoms of radioactive beryllium-10 (^{10}Be) in the clay. ^{10}Be is an isotope of the element beryllium with ten protons and neutrons and is formed when cosmic radiation from space breaks up oxygen and nitrogen atoms in the atmosphere. Since cosmic rays bombard the Earth in a steady rain, the amount of ^{10}Be in the clay should tell how many years the clay layer took to form.

Unfortunately, the ^{10}Be method didn't work. The isotope's half-life is just a little too short, so that hardly any of it remained in 65-million-year-old clay. But the failure set Luie thinking: Was there anything else from space that would have ended up in the clay? What about micrometeorites? These tiny dust grains are left over from the constant rain of small meteors that vaporize harmlessly when they hit Earth's atmosphere. Micrometeorites are constantly settling to Earth. If their number could be counted in the mysterious clay layer, they could provide the clue. But how to count them? Many of them are too small to see even with a microscope.

While Luie was looking for a way to use nuclear physics, his specialty, to count the micrometeorites, he realized something important. Platinum, gold, and certain other heavy metal elements are ten thousand times more common in meteorites than in the Earth's crust. While the Earth was hot, molten rock, gold, platinum, and their relatives alloyed with iron, and gravity pulled them to Earth's core, where they remain today beyond reach of the most audacious prospectors. Luie was able to show that *most* of the platinum-group elements in sedimentary rock and clay in fact came from meteorites. But there would still be only a few parts per billion of these elements. How could he find and measure such a tiny amount?

After considering and discarding several other techniques, Luie hit on a way to identify the rare element iridium in the clay, using a method called neutron-activation analysis. Few people have ever heard of iridium, but it is used by jewelers to make platinum jewelry harder and also sometimes in ball-point pen tips for durability. It is, with osmium, the densest of all elements (22.5 times denser than water, or about twice as dense as lead).

Luie, Walter, and neutron-activation experts Frank Asaro and Helen Michel found iridium in the clay layer, about one part in a billion. In the limestone above and below the clay layer, there was no detectable iridium. But as it happened, there was too much iridium in the clay layer for micrometeorites to be the main source of it.

After several months Luie concluded—erroneously, as it turned out—that the iridium had come from a supernova, a star that reaches the end of its life and explodes. The powerful blast wave of a supernova explosion creates temperatures of more than a hundred million degrees. Under these extreme conditions, unmatched anywhere else in the present universe, heavy elements like lead, gold, and iridium are created and flung out into space. Supernovae are rare, about one per galaxy every fifty years, but over the immense age of our Milky Way galaxy, some heavy element material would have been spread throughout its volume.

The idea that a supernova killed the dinosaurs was not new—physicist Mal Ruderman had proposed it years before. If a supernova erupted close enough to Earth, the blast wave could have blown away the atmosphere and killed life outright with deadly high temperatures. Even if the supernova wasn't so close, intense nuclear radiation from it could have finished off most living species.

A supernova would also create plutonium, Luie realized, the notoriously radioactive element used to make nuclear weapons. Plutonium is almost completely missing from the Earth's crust—most of the world's supply has been produced from uranium in nuclear reactors. Although plutonium decays, Luis knew that if a supernova had injected it into Earth's atmosphere 65 million years ago, some should still be left. Now the question was, did the 65-million-year-old clay layer contain plutonium as well as iridium? Frank Asaro and Helen Michel had to perform extraordinary feats of radiochemistry to find out. The answer turned out to be negative. The supernova hypothesis was dead.

But Luie still felt that the iridium had come from space, and he was determined to find its source. A theoretical astronomer, Chris McKee, had suggested to Luie that an asteroid hitting the ocean could cause a tsunami or giant wave. Maybe the wave itself had killed the dinosaurs. But how could waves have reached the middle of continents, thousands of feet above sea level? And how could a wave, no matter how large, kill off sea creatures worldwide?

Fred Hoyle had written a science-fiction story featuring a dust cloud that blocks sunlight and brings freezing temperatures even to tropical parts of the Earth. Luie considered the possibility that an iridium-rich asteroid could hit the Earth, blast out a huge crater, and kick up a lot of dust. Iridium would be carried all over the world by high-altitude winds, then fall back to Earth and be incorporated in the worldwide sedimentary layer.

The impressive Meteor crater in Arizona is 1.2 kilometers across (just under a mile), 200 meters deep, and between 25,000 and 50,000 years old. It was blasted out by an iron projectile (whose remnants have been found) 50 meters across that slammed into the desert at a speed of about 11 kilometers per second (25,000 miles per hour). There are much larger craters on Earth—nearly a hundred are known. But they have been partially or totally covered by erosion and other geological activity. In southern Germany, the 25-kilometer-wide Ries crater has been connected to the impact of an asteroid some 15 million years ago. The huge Manicouagan ring in Quebec, now a lake, marks an impact crater about 100 kilometers across and 210 million years old. It wasn't even noticed until the dam that created the lake was built. In Siberia, the Popagai crater is about the same size and 37 million years old. Off Nova Scotia lies a 45-kilometer submerged crater about 50 million years old. In Iowa the buried Manson structure is 35 kilometers in diameter. The Vredefort impact structure in South Africa is 140 kilometers in diameter. And half submerged off the Yucatán peninsula of Mexico (although not known to Luie or other scientists at the time) is the largest crater so far recognized on Earth, more than 170 kilometers across and accurately dated at 65 million years old.

With few exceptions, large Earth craters are not obvious from an airplane or even from space because of the effects of mountain building and erosion. Scientists must detect them by studying local variations in the Earth's gravity or magnetism or other geological anomalies. Craters are far more common on other planets and moons of the solar system than on Earth—and much easier to see. The rugged surface of the Moon is pockmarked with more than thirty thousand craters of all sizes. With their symmetrical raised rims and dimple at the center, many of the smaller and medium-sized ones are textbook illustrations of impact physics. More difficult to explain in detail are the larger craters; forty of these lunar basins are more than 300 kilometers in diameter, with complex multiringed rims. One, the Procellarum basin, is more than *3,000* kilometers in diameter. Scientists have little doubt that these lunar scars mark the sites of ancient collisions with asteroids or comets.

Stark radar images of the surface of Venus, captured by the orbiting Magellan satellite, reveal numerous impact craters as well. Close study of these remarkable images even shows from which direction the asteroid or comet arrived. The largest of these impressive craters, Mead crater, has a diameter of 280 kilometers, larger than any crater known on Earth. The curious backward rotation of Venus is believed by some to be the result of a gigantic impact whose "footprints" were erased by later asteroid collisions.

Missions to Mars and Mercury, too, have revealed heavily cratered landscapes dotted with multiringed basins explicable only by cataclysmic impacts. Spectacular pictures of the moons of Jupiter and Saturn taken by the Pioneer and Voyager spacecraft and the breathtaking flyby images of asteroids Ida and Gaspra also reveal intense cratering.

As he closed in on the iridium mystery, Luie Alvarez read articles on asteroids with Earth-crossing orbits, the so-called Apollo objects. He quickly determined that the largest one most likely to hit our planet in a hundred-million-year interval would be about 5 (maybe 10) kilometers in diameter. He found it somewhat less likely that a comet with a core of that diameter would hit Earth every hundred million years. Collisions with smaller objects are more likely, since there are many more small asteroids than large ones.

Becoming more and more excited about the asteroid hypothesis, Luie set about calculating what effects a collision with Earth would have. (Simple computations like those Luie made are what physicists are fond of calling "back of the envelope" calculations. A physicist having dinner in a restaurant will often scrawl them on a napkin or even a matchbook.)

If an asteroid crashes into Earth, Luie figured, its relative velocity could easily be 30 kilometers per second, the same speed of our planet around the Sun and about thirty times the speed of a high-powered rifle bullet. Greater impact speeds are ruled out for asteroids (but not comets) because all known asteroids circle the Sun in the same direction as the Earth. Could such an impact knock Earth out of its orbit? The answer depends on the asteroid's momentum, or its mass times its velocity. Since the asteroid and the Earth have roughly the same velocity, the main issue is that of mass. How many times more massive than the asteroid is Earth? The Earth's diameter is about 12,800 kilometers. This is over two thousand times larger than a 5-kilometer asteroid. Assuming the bodies have a similar density—not unreasonable since both are made of rock—the relative mass will go as the cube of the diameter ratio, or more than two thousand times two thousand times two thousand. So the momentum of the asteroid is about one ten-billionth that of the Earth. Its smashing into Earth could alter our orbit by no more than one ten-billionth of the 93-million-mile distance from Earth to the Sun, or about 50 feet. Not to worry—this would have no significant effect.

But as every high school physics student or gun enthusiast knows, momentum is only part of the story in a collision. A moving object also carries *kinetic energy.* And something traveling up to thirty times faster than a rifle bullet has potential to make big trouble in a collision. Kinetic energy increases as the square of velocity, so the energy of each gram of asteroid traveling 30 kilometers per second would be about nine hundred times the energy of each gram of speeding bullet. In the spirit of a back-of-the-envelope calculation, we'll call this number 1,000. This energy all comes from the explosive, which is about 10 percent of the mass of the bullet. So to compare the energy of an asteroid with that of explosive (like gunpowder or TNT), we just have to multiply the asteroid mass by 10 percent of 1,000, or 100. So every ton of asteroid carries the energy of 100 tons of TNT. The mass of a 5-kilometer asteroid is about 1 million megatons (10^{15} kilograms). So its impact with Earth will release the energy of a hundred million megatons of TNT. That's a hundred thousand times greater than the combined nuclear arsenals of all the nations on earth.

Never in recorded history had that much energy been dumped at one spot on the Earth's surface, Luie knew. What would the effect have been? He looked up published estimates of the energy needed to make impact craters on the moon. He researched the largest nuclear explosions in the U.S. test program and found out how big the craters were. His conclusions were astonishing, some would say frightening. A 5-kilometer asteroid would make a crater nearly a hundred miles across. It would produce temperatures of more than a million degrees, vaporizing surrounding rock and melting far more. And it would throw enough material into the atmosphere to block sunlight. That was it, Luie concluded. Darkness from such impacts had led to the death of plants and later, animal life, including the dinosaurs.

Luie was familiar with the sky-darkening effect of dust in the atmosphere from studying the enormous explosion of the Krakatoa volcano in the South Pacific in 1883. It had blasted dust and rock more than thirty miles into the air. Dust spread worldwide and for years caused spectacular reddened sunsets thousands of miles away. It took months for most of the dust to settle. Temperatures dropped worldwide by at least half a degree Celsius. In 1991 the eruption of Mount Pinatubo in the Philippines reddened sunsets on the Pacific Coast for over a year and produced a slight worldwide cooling.

Before Luie could have confidence in the impact theory, he needed to check the amount of iridium it would lay down. He assumed the fraction of the asteroid that was iridium was similar to that in meteorites, about half a part per million. He calculated the total mass of iridium and how much would be found in the 65-million-year-old clay layer if some reasonable fraction of the asteroid's entire mass were spread out over the world. About four-fifths of the asteroid had remained at the impact site, Luie figured; the rest had blasted into space, then rained down onto the Earth. The iridium concentration came out right! Fifty thousand tons of iridium had been scattered around the Earth. What was more, Luie could account for the entire clay layer, not just the iridium, as coming from asteroid debris and the greater amount of rock kicked up from the crater.

The more Luie and Walter checked the impact theory, the more confident they became. The theory made numerous predictions that couldn't be tested immediately but that would later allow the theory to be confirmed or rejected decisively, either by the Berkeley group or by others. (In science, a theory that cannot be ruled out is useless; one that makes many testable predictions is wonderful.) The 65-million-year-old clay layer had to be rich in iridium everywhere in the world. Chemically, the clay should be the same everywhere. If other mass extinctions were caused by asteroids, there should be evidence of impacts associated with them. And somewhere in the world there had to be a 65-million-year-old crater at least a hundred miles across.

Within a few years all these predictions were confirmed. When the iridium discovery paper, entitled "Extraterrestrial Cause for the Cretacious-Tertiary Extinction" was published, Luie's physicist colleagues applauded. Geologists and astronomers were sympathetic. But paleontologists, whose "official" business it was to explain things like why the dinosaurs died off, thought the Alvarez discovery had all the earmarks of a crackpot theory. With its killer asteroid from outer space, Luie's theory seemed wild and irresponsible. Had its principal author not been a Nobel prize winner, they could have just ignored it. Instead, what followed was one of the most furious and fascinating controversies in the history of science.

CHAPTER FOUR

Controversy

"A nutty theory of pseudoscientists posing as paleontologists"—so wrote one famous paleontologist in a letter to *The New York Times* about the Alvarez impact theory. "Alvarez is so contaminated with iridium that he glows in the dark," ran a favorite joke of graduate students in the Berkeley paleontology department. But the real joke was that these students did not even know that iridium is not radioactive. Still, why the insults? Was this any way to talk about a man who some considered the world's most distinguished living experimental physicist?

In 1980, most paleontologists believed that mass extinctions had been caused by gradual climate change. According to their favorite scenario, dinosaurs vanished as the shallow inland sea that covered much of the United States receded, generating major and complex alterations of climate. Gospel among paleontologists held that mass extinctions do not have a single simple cause. Furthermore, many of the attempts by outsiders—nonpaleontologists—to explain the disappearance of the dinosaurs were simply nutty.

But the main reason that the concept of a sudden catastrophe aroused so much fury among paleontologists was that it violated the dominant philosophy of earth science that is drilled into every freshman geology student: *uniformitarianism.* According to this view, important changes in the history of our planet happen slowly, with the possible exception of volcanic eruptions. The notion of gradualism was originally a daring refutation of the traditional view espoused in the Bible. After all, it is still only about 150 years since geologists broke free of the biblical story that the world was created six thousand years ago in a mere six days. The Bible is filled with catastrophes. For example, Noah's flood threatened to cause mass extinction. Only by building an ark did he save the large animal species.

It had taken geologists and paleontologists decades, even centuries, of patient sleuthing and furious debate to establish that geological and biological evolution on the Earth have spanned billions of years—and that most of it can be explained without catastrophes. Only in the 1950s had geologists applied radioisotope dating to prove that the Earth is 4.5 billion years old. In 1835 Charles Lyell, the founder of modern geology and a champion of gradualism, had attacked notions of "the sudden annihilation of whole races of plants and animals" as being in "the ancient spirit of speculation": that is, not scientific.

Mountains, we know now, are built over millions of years by uplift, then worn down by erosion of wind and water over more millions of years. Much of the rock we see is sedimentary, having been slowly deposited over aeons at the bottom of the sea. Streams and rivers wear slowly away at their banks, gradually changing course and sometimes gouging out canyons. (A resident of Missouri whose home was submerged when the Mississippi River changed course during the great 1993 flood might disagree with the word *slowly.*)

So thoroughly committed to the doctrines of gradualism and uniformitarianism were geologists of the early and mid-twentieth century that for five decades they bitterly fought Alfred Wegener's brilliant theory of continental drift. Wegener argued that continents, made of less dense material than the underlying molten magma, float on top of it and move slowly toward and away from each other. Plate tectonics, the technical name for these processes, we now understand as the primary cause of earthquakes and volcanic eruptions. In fact, the study of plate tectonics has become central to modern geology. Curiously, considering the furor it created, the revolutionary idea of continental drift didn't really even violate gradualism, since the motion of the plates is measured only in centimeters per year.

In astronomy, the role of catastrophe had come to be accepted by 1980. The most widely accepted theory of the origin of the moon involves a collision of Earth with another planet. So much for gradualism! But outside astronomy, scientists hate to invoke rare isolated events to explain their data, since such events are so much more difficult to study than forces that act gradually. The random, the unpredictable, and the chaotic arouse distaste in many scientists, who are apt to lump them together with the pseudoscience of flying saucers, ghosts, extrasensory perception, and mental spoon-bending.

In the 1950s and 1960s, a medical doctor of Russian extraction named Immanuel Velikovsky gave the idea of planetary impacts a bad name—at least in the world of orthodox science. Overreacting to the excessive gradualism of geologists, he presented dramatic accounts of destructive impacts in his wildly popular books *Worlds in Collision, Ages in Chaos,* and *Earth in Upheaval.* According to Velikovsky, these collisions happened not billions or millions of years ago but in historic times, and Middle Eastern peoples had actually observed and recorded their effects. His ideas were based not on geological observations or mathematical computations but on the study (most scholars would say the misreading) of ancient texts and myths. His apocalyptic vision had great appeal to the scientifically uneducated, but it showed little respect for the laws of physics: In his books, planets change their orbits—in blatant violation of the laws of mechanics—and go crashing into other planets.

Given all this background, it is not surprising that geologists and paleontologists fiercely resisted the impact theory and the notion of a "K-T catastrophe." (*K-T* is short for "Cretaceous-Tertiary," signifying the boundary between these two major geological periods.) At first the two sides merely talked past each other, without necessarily listening to the details of each other's cases. Paleontologists "knew" that dinosaurs had died out over a period of millions of years, while Luis Alvarez "knew" that only an extraterrestrial event could explain the iridium. Over the years, while Luie attempted to rebut his many critics, his team, led by son Walter, continued to collect rocks and analyze them. So compelling was the debate that hundreds of geologists reoriented their careers to participate.

By the mid-1980s, iridium had been found at eighty sites around the world. (Now there are more than a hundred sites, and some three thousand scientific papers have been published related to the K-T catastrophe.) The iridium-bearing clay layer was found to be chemically similar in Denmark, Italy, Montana, under the North Pacific Ocean, and wherever else geologists came across it.

Some skeptics argued that the iridium could have been laid down in the ocean as a result of a shift in oceanic chemistry. But in 1984, a U.S. Geological Survey team led by Carl Orth found a high level of iridium in a deposit that had never been under the sea. In the same sample the ratio of pollen to fern spores dropped abruptly just as the iridium jumped. This showed that plant life was affected at the same time as animal life.

Geologists associated with the Alvarez team found unusual little "spherules" in the dark clay layer but not in the surrounding limestone. The little spheres, known as microtektites, are fossilized glassy beads formed in the intense heat of an impact, when droplets of molten rock fly off and then solidify again when they cool. But microtektites can also be formed by a powerful volcanic eruption. The Alvarez team, it seemed, could not rule out a volcanic explanation for the mass extinction of 65 million years ago.

Another telltale finding in the clay layer excluded the volcanic explanation. When common quartz is squeezed by enormous pressure, such as under the impact of a meteorite or a nuclear bomb, its crystals form a unique layered structure. Crystals resembling these so-called shocked quartz crystals were found in the 65-million-year-old clay layer. Such crystals are not found in volcanic debris. Not even the most powerful volcanic eruption can build up as much pressure as an asteroid or comet impact.

At the Brazos River site in Texas and at sites around the Caribbean, the boundary clay layer contains jumbled rocks, which geologists think were dumped there by a tsunami wave. There are also glassy spherules with microtektite characteristics. Some glassy material recovered from outcrops in Haiti and elsewhere in the Caribbean has been dated. Its age: 65 million years.

Despite this seemingly overwhelming evidence in favor of the impact theory, some geologists and many paleontologists still dispute it. A small band of geologists continue to maintain that a volcanic eruption was the cause of the mass extinction of 65 million years ago. Could any amount of evidence convince them?

Some paleontologists cite the fact that crocodiles and turtles—which are presumably as vulnerable to cold as dinosaurs—survived the mass extinction, as evidence against the impact theory. However, they do not attempt to explain this fact themselves. In any case, the impact hypothesis does not hold that all life was wiped out worldwide; nor does it deny that other causes of extinction were operating at the same time.

Some skeptics argued that there could have been no sudden die-off of dinosaurs since dinosaur fossils have a vertical spread. In fact, the last appearance of some species is well below the 65-million-year-old clay layer. Dale Russell, a well-known Canadian paleontologist, showed that it is statistically possible to account for the vertical spread because dinosaur fossils are rare (nearly all are found in North America), and the last-well-preserved skeleton of a particular species could have been fossilized millions of years before the impact.

There is some indication that the extinctions were "stepped"—that is, took place in several stages over hundreds of thousands or even millions of years. This possibility cannot be excluded because these are gaps in the fossil record, but it doesn't really contradict the idea of at least one very destructive impact. There could have been more than one impact—or the effects on some species may have been long delayed.

The Alvarez hypothesis got a boost in 1985, when University of Chicago chemists Edward Anders, Wendy Wolbach, and Roy Lewis discovered soot in the clay layer. Soot is basically carbon, such as is formed by burning wood. Like the iridium, the soot was found worldwide. By itself, soot is not really evidence of an impact. But given the other overwhelming evidence that a major impact rocked the Earth 65 million years ago, the soot is a direct indication of the impact's biological effect. Scientists found so much of it, in fact, they concluded that many of world's forests and grasslands must have burned at once—at least 10 percent of all the living mass on the Earth must have gone up in flames.

Scientists are still arguing over which effects of the impact did the most damage to life on Earth. Many doubt that dust and soot alone could have blocked photosynthesis long enough to cause mass extinction, especially on land. Larger dust particles drop out of the atmosphere in a matter of days, and even the smallest such particles would have settled in six months. For now, some atmospheric scientists are placing their bets on the sulfuric acid haze, which could have lasted a century, as the effect most destructive of life. Other scientists believe the immediate effects of the blast and heat were the most destructive, at least for life on land.

There have been many mass die-offs of marine life in the last 600 million years, perhaps as many as twelve. About 250 million years ago, at the end of the so called Permian era, more than 90 percent of all species of fossil sea creatures vanished in a mass extinction that overshadows more recent ones. Another great mass extinction dates from 365 million years ago, at the end of the Devonian epoch. Do these mass extinctions have extraterrestrial causes too? Geologists have found excesses of iridium, the telltale sign of an asteroid or comet crash, and little glassy spherules in association with the latter two extinctions, but in neither case is the iridium excess as spectacular and widespread as in the case of the K-T catastrophe.

Smaller mass extinctions have also threatened life in the last 50 million years. For two of these catastrophes, which happened about 38 million and 12 million years ago, boundary layers rich in iridium were found, and glassy spherules too.

Why are some extinctions clearly associated with iridium and some not? One possibility is that for many extinctions, especially the most ancient, we have yet to locate the iridium-containing boundary layer. The oldest extinctions are much harder to study than the K-T catastrophe since geological processes of uplift and erosion have had more time to smear, smash, and jumble the fossil record. Another possibility is that some catastrophies are caused by grazing impacts, in which nearly all the comet core mass bounces back out into space. Finally, it is also conceivable that some mass extinctions were caused by impacts and some by other causes, such as volcanic activity.

Most scientists have now accepted that an impact was responsible for the catastrophe of 65 million years ago. By the late 1980s, the only missing piece in the Alvarez argument was the crater. After ten years of searching, it was found, but unfortunately Luis Alvarez did not live to see the confirmation of his "nutty" theory. He died in 1988.

CHAPTER FIVE

The Smoking Gun

If an asteroid or comet killed the dinosaurs, it must have left behind a crater. As intense debate swirled around the Alvarez discoveries, geologists began a worldwide search. At first the clues they found were minimal. Of the hundred or so known impact sites, only a few were dated at or near 65 million years, but these were too small to account for a great die-off. A crash big enough to loft sufficient dust into the atmosphere to blot out the Sun for months would have made a crater 150 to 200 kilometers in diameter. Despite this huge size, the odds for finding the crater weren't particularly good. If the crater were on land, erosion could have hidden it completely. If it were initially under the sea or if it had been inundated after the impact, thick layers of sediment could have buried it.

Even today three-quarters of the Earth's surface is ocean, much of it kilometers deep and excruciatingly difficult to explore. During the warm Cretaceous era, when dinosaurs ruled, even more of our planet was water-cloaked. Furthermore, the K-T extinctions affected ocean life more dramatically than land species; a greater percentage of them survived. A lawyer might argue that circumstantial evidence points overwhelmingly to an ocean impact. But 65 million years of movement by Earth's crustal plates could have obliterated the crater altogether, especially if it were in the ocean. According to the best estimates, as the plates have ground over one another in the epochs since the great die-off, most of the ocean floor has rejoined Earth's mantle. One clue did suggest that the impact site lay at least in proximity to land, if not on it: the tiny grains of shocked quartz found throughout the world in the clay layer, along with abundant iridium from an extraterrestrial projectile. Far from the continental margins, the deep ocean bottom contains little or no quartz. Meteorites, too, contain little quartz. So the presence of quartz crystals in the clay layer suggests strongly that the crater must have been on or near land, where quartz is common.

In the 1950s, geophysicists of the Mexican national oil monopoly, Pemex, were hunting for oil in southern Mexico. They were looking for irregularities in the Earth's gravity, which can reveal unusual structures of rock deep underground or beneath the sea floor. Along the northern coast of the Yucatán peninsula, famed for its Mayan pyramids, they located some potentially interesting formations more than a kilometer down. Wells were drilled, but no oil was found. Some of the wells cut across rock that clearly had been melted. At first, no one suspected that a powerful impact was responsible for the melting; much less was known about terrestrial impact craters in those days. Mistakenly, Pemex geologists classified the formations as volcanic domes.

By the 1970s, petroleum geologists had a valuable new tool for discovering oil. By flying extremely sensitive magnetic detectors in airplanes, they could look for telltale variations in the Earth's magnetic field. In 1978 Pemex decided to survey the Yucatán again and hired a Texas company, Western Geophysical, to carry out the search. Glen Penfield, a young geophysicist working for the company, found a strange pattern in the magnetic data. Beneath shallow waters to the north of the Yucatán peninsula lay a great curved swath of something—rock different from the limestone sediments that dominate Yucatecan geology. On a Pemex gravity map of the region, Penfield found another arc, this one mostly on land but curving the other way from the magnetic map. Combined, the two patterns formed a nearly perfect circle. Penfield excitedly recognized the impact crater, part hidden under the Caribbean Sea, part buried deep under Yucatecan soil.

Pemex, protective of what any oil company would consider proprietary information, at first refused to permit publication of the data. In 1981, as the Alvarez iridium discovery was redirecting many geological careers, however, Penfield and his Pemex boss, Antonio Camargo, finally got permission to present their data publicly. It may seem surprising now, but the geological community did not exactly cheer Penfield's discovery when they heard of it. In fact, they made hardly any response at all. Perhaps this curious development resulted from the fact that Penfield and Camargo were outsiders to academic geology and did not present their findings in the journals and conferences where the impact controversy was then raging. In any event, not until a decade later was the Yucatán crater—now known by the name Chicxulub, after the sleepy fishing harbor at its center—accepted as the unchallenged site of the great impact.

During the 1980s, independent evidence built up pointing to the Caribbean at the impact site. All attempts to link known land craters with the K-T boundary clay layer and its iridium excess had failed. But around the Caribbean, geologists discovered jumbled deposits of coarse rock, up to several meters thick, that looked as if they had been thrown down by a series of giant tsunami waves. The tsunami deposits turned up in Alabama, Texas, and Mississippi, in Cuba, in the Mexican states of Chiapas, Nuevo León, Tamaulipas, and Veracruz, and even in deep sea cores recovered off Florida and Haiti. These strange deposits were very rich in impact spherules and shocked quartz, and they were closely associated with the K-T boundary in the fossil record—precisely where mass extinction had occurred.

As for iridium, chaotic tsunami deposits around the Caribbean were found to be topped by layers strongly enriched with the telltale element. Unlike the thin (1 centimeter) clay layer from Gubbio, Italy, these thick deposits had several different iridium-rich layers. This multiplicity could have had several causes. Some iridium might have been deposited right after the asteroid or comet hit, and some hours or days later. When the tsunami waves struck, they might have jumbled the previously deposited layers and created a new layer from iridium-containing material of their own. Afterward, more iridium-containing dust might have fallen from the darkened sky.

Alan Hildebrand, a young geologist from the University of Arizona, was among the first to recognize the significance of the tsunami deposits. In greenish clay exposed near the Haitian village of Beloc, he found all the evidence of material ejected from a faraway explosion. The impact had occurred within or near the Caribbean basin, he judged. By chance, he learned of Penfield's earlier work. Most of the Pemex drill cores had been lost, but a few samples turned up. When Hildebrand, Penfield, and their colleagues analyzed the precious samples, they found abundant shocked quartz—the unambiguous signature of a violent impact.

The Chicxulub crater drill samples, as well as the glassy spherules found hundreds of miles away in sediments around the Caribbean, have been dated by radioactive-isotope methods. The ages of the crater samples and the glassy spherules agree perfectly: 65.0 million years.

Yet another kind of data supports the identification of the Chicxulub crater with the K-T impact. Scattered about the Yucatán are monster limestone sinkholes called cenotes. In Mayan times, these water-lined pits, along with the pyramids, were sites for rituals of human sacrifice. Today some of them are used as wells, and one cenote in the capital, Mérida, has been converted into a romantic restaurant grotto. These sink holes appear in satellite images, which otherwise show no trace of the Chicxulub crater, lying in a great arc centered on the fishing village of Puerto Chicxulub. The arc corresponds to the outline of the crater rim as determined from gravity and magnetic data; its diameter is 170 kilometers.

The Chicxulub crater is the largest crater known on the Earth. Like other very large craters, it is made up of three major zones. The middle zone, about 90 kilometers in diameter and several kilometers deep, was formed when the initial blast created a tremendous cavity, which promptly collapsed. It contains most of the material melted in the impact. Hildebrand and his collaborators estimate that the volume of rock melted was about 20,000 cubic kilometers. At the center of the crater is an uplifted region 40 kilometers in diameter, formed from the violent rebound of the cavity floor. Finally, the vast inwardly tilted outer zone, 170 kilometers across, was also blasted out during the collapse of the initial cavity. Some geologists claim to have found evidence for faint rings outside Chicxulub's prominent 170-kilometer-wide edge; the largest of these may be as much as 300 kilometers across. Even if these ring estimates prove incorrect, the crater covers a vast area. Any of the world's largest metropolitan areas—New York, Los Angeles, London, Mexico City, São Paulo—could fit neatly inside, with all their suburbs and much of the surrounding countryside.

In the vast ring around the crater, geologists continue to discover material blasted into space by the crash—so-called ballistic ejecta. The farther they go from the site, the fewer ejecta they find. Rock in the Yucatán contains much sulfur, which supports the idea that sulfuric acid rain made the impact especially deadly. The thin, nearly uniform layer of clay and soot found elsewhere in the world in 65-million-year-old sediments is missing around the Chicxulub crater. However, the chaotic deposits rich in iridium found all around the Caribbean region show the effects of huge tsunami waves radiating from the crash site.

Geologists are attempting to model the impact that blasted out the Chicxulub crater in as much detail as possible. Was the invading object an asteroid or a comet? Did it come nearly straight down or obliquely? How big was it? By comparing Chicxulub with radar images of well-defined craters on Venus and other planets, P. H. Schultz of Brown University has come to some intriguing conclusions. On Venus, oblique impacts are easily recognized by the patterns of ejected rocks deposited downrange. The craters are dug out more deeply in the direction the object came from (uprange) and shallower in the opposite direction (downrange). The inner ring of Chicxulub is open to the northwest; the outer ring seems to have a gap in this direction; and magnetic patterns in the crater point to the northwest. From these and other data, Schultz concludes that the projectile zoomed in from the southeast at about 30 degrees above the horizontal. If it was between 10 and 15 kilometers in diameter, its speed would have been between 30 and 15 kilometers per second. (A smaller size requires a larger speed to produce the observed damage; the *average* impact velocity of meteorites is about 17 kilometers per second.)

Such a slanted trajectory would have produced a horrific cloud of hot vapor that blasted its way toward the northwest. A 650-kilometer-wide firestorm with hurricanelike winds would have engulfed much of the Gulf of Mexico and boiled its upper waters. The blast would have driven molten and solid rock toward what is now the western United States at supersonic speeds, bringing about stupendous and almost immediate destruction of habitat and animal life. Still more energy released by the crash would have been concentrated in a nearly symmetrical cloud of debris thrust upward from the central impact zone. After the initial battering and firestorm, much of North America would have suffered a second bombardment of fiery missiles and choking sulfurous gases.

Hildebrand and his colleagues' analysis favors a steeper entry angle for the incoming projectile than Schultz's. They find that only some of the melted rock in the central crater holds excess iridium. Comets are expected to contain less iridium than asteroids of similar energy. Comparing the amount of iridium deposited worldwide with the mass of the projectile needed to blast out Chicxulub, Walter Alvarez, Hildebrand, and other scientists have concluded that the projectile was probably a comet. Only if most of the iridium ricocheted back into space could the crater have been made by a typical asteroid.

Could another Chicxulub happen in the future? How many asteroids and comets are streaking around the solar system in orbits that could menace Earth? Do they hit our planet randomly, or do they cluster in meteor showers or comet storms? What happens when smaller comets hit Earth, if indeed they do? Like good military commanders, we inhabitants of this planet must know our enemies.

CHAPTER SIX

Asteroids

On August 28, 1993, a most extraordinary event took place. The Galileo spacecraft acquired close-up images of a rocky, irregularly shaped object hundreds of millions of kilometers from Earth. Scarred by thousands of craters, the object resembles at first glance nothing more than a pockmarked potato. It is the asteroid Ida, no less than 52 kilometers across. Near Ida, Galileo discerned its tiny companion, Dactyl, measuring only 1.6 kilometers long. The discovery of Dactyl, also peppered with craters, confirmed what many amateur asteroid watchers had long claimed: These vagabond meteoritic rocks sometimes come in pairs.

There is no hard and fast distinction between a meteor and an asteroid. Both are basically rocks that come in a continuum of sizes, from a grain of sand almost to a respectable planet. A meteor, or so-called shooting star, comes to a fiery but harmless end in Earth's atmosphere. A meteorite is a meteor that survives its fall to Earth; there is a 34-ton example at the American Museum of Natural History in New York and an even bigger one on display in Africa.

On its way to probe Jupiter, Galileo encountered asteroids, or obscure miniature planets, in the belt where most of them orbit the Sun, between Mars and the giant gas planet. In October 1991, Galileo passed close by Gaspra, a body one-third the size of Ida, and radioed home our first clear picture of an asteroid. For astronomers, the Galileo flybys gave asteroids a new reality. No longer were they mere pinpoints of light in the sky, dimly discerned on astronomical photographs. Looking at these amazing images with the Chicxulub crater fresh in mind, one couldn't help imagining the worst-case scenario: one of these cosmic potatoes smashing into Earth at twenty to forty times the speed of a rifle bullet.

Astronomers currently estimate that asteroids more than 1 kilometer in diameter number between a hundred thousand and a million. Rocky bodies with a smaller diameter but still more than a few meters across probably number in the billions. By comparison, the average meteor is smaller than a grain of sand. The ones that make an awesome fireball lasting several seconds or more are usually about the size of a pea. Most meteorites, rocks large enough to survive reentry, are no larger than your fist. Rarely do they smash through roofs or hit humans. In the cosmic shooting gallery where humans are among the targets, asteroids and comets are the bullets. How many of them are aimed at us, and how likely is a dangerous hit?

It would be comforting to know that all potentially dangerous asteroids had been discovered and were continuously tracked by vigilant astronomers. Then we would not have to live in fear of them. But the truth is that more potentially lethal undiscovered asteroids wander across our path than known ones. And although astronomers first detected asteroids nearly two hundred years ago, only within the last few years have they grasped how some escape from the main asteroid belt onto rogue orbits that intersect our own orbit around the Sun.

These rogue asteroids are called Earth-crossers. The first such potentially deadly moonlet to be discovered, named Apollo, was spotted in 1932 but was "lost" soon thereafter. In 1936, Adonis, which we now believe is just under a kilometer in size, came within 2 million kilometers of Earth—about five times the distance of the Moon. In 1937, Hermes, a true "Earth-grazer," zipped past us at a mere distance of twice the Moon's. We learned of another such close call in 1989, but only *after* the 1-kilometer alien had passed the Earth. Had this little caller shown up six hours earlier, it might have dealt us a million-megaton civilization-destroying blast. In 1991, a baby asteroid in the 10-meter size range approached even closer, to within half the distance to the moon.

By the late 1950s, astronomers knew of eight Earth-crossers, but they had lost track of most of them. Some, like Apollo, were rediscovered by accident—Apollo came within 9 million kilometers of Earth in 1980 and again in 1982. Since few observers seemed interested in cataloging them, geologist-turned-astronomer Eugene Shoemaker (who co-discovered comet Shoemaker-Levy 9) decided to make this his specialty. Today, thanks mostly to Shoemaker, his wife Carolyn, and their co-workers, we know of about eighty Earth-crossers. A few dozen have paths that could potentially intersect the orbit of Mars as well; these are called Aten or Amor asteroids. Spending much of their time beyond the orbit of Mars, they can be strongly affected by Jupiter's gravity. Strangest of all is Icarus. This mountain-sized rock about 2 kilometers across has a stretched-out orbit that flings it way beyond Mars on one end (its apogee) and far closer to the Sun than even Mercury ever comes on the other (its perigee). At its nearest approach to the Sun, it may glow red hot. In its present orbit, it can get within only sixteen times the Earth-Moon distance, so it does not seem to be a threat.

The big brutes among Earth-crossers are about 10 or 20 kilometers in diameter, around the same size as the killer asteroid (or comet) that did in the dinosaurs. Clearly such objects should be watched—carefully! Eugene Shoemaker and other astronomers estimate that there are more than two thousand Earth-crossing asteroids of kilometer size or greater, of which about 90 percent have yet to be discovered. Any one of them could make a catastrophic bang. A 1-kilometer asteroid would blast out a crater about 13 kilometers in diameter—big enough to swallow up a city the size of San Francisco. The area it devastated would be many times larger. Sun-blocking dust and noxious mists propelled upward might cause worldwide death from starvation if not mass extinction. Even more shocking, *most* of these asteroids probably are destined to collide with Earth—someday. Those that do not, and that also avoid a terminal impact with Mars or Venus, will eventually be ejected from the solar system by their gravitational interactions with the planets, especially Jupiter. A few will collide with each other and shatter. But there's little cause for optimism about the long term: As Earth-crossers go out of circulation, others arrive from the main asteroid belt to take their place. Estimates of the collision rate between asteroids of a kilometer or more and Earth vary from about once every 250,000 years to roughly once every 2 million years. This is not such a low rate that you—your insurance company—can afford to ignore it.

What about the largest asteroids—do they pose a threat too? For the most part, barring some unforeseen revelation in celestial mechanics, they seem not to. The grander miniplanets seem to whirl around the Sun in stable orbits, permanently trapped between Mars and Jupiter. The monarch of the asteroid belt is Ceres, with an estimated diameter of 900 to 1,000 kilometers, about one-third the size of the Moon. Next biggest are Pallas and Vesta, each between 500 and 600 kilometers in diameter. Another thirty measure more than 200 kilometers, while more than two hundred measure larger than 100 kilometers across. All of these are substantial little planets about which whole books (or CD-ROMs) will someday be written. In all, the orbits of some three thousand asteroids are accurately known. There have been thousands of mere "sightings," but to be credited with a discovery, an astronomer must track the asteroid long enough to accurately determine its orbit. At least a hundred thousand can be imaged from Earth with currently available technology.

Asteroids of a certain size are relatively easy to spot. Like planets, they show up as linear trails on long-exposure photographs of the sky. Some are found by accident—for example, when astromers scan galaxy images for supernovae. Amateur astronomers using modest-sized telescopes have discovered hundreds of asteroids. As computerized remote astronomy spreads to schools, students are likely to find many more. But most of the asteroids being discovered today are found by professional asteroid hunters, especially by the Spacewatch program of the University of Arizona. As we will see, astronomers are planning to boost the discovery rate so that they can eventually find most of the Earth-crossers.

Measuring an asteroid's size is crucial to gauging its potential for destruction. The best measurement method astronomers have, however, is only indirect, making use of the amount of light reflected by the body (its observed brightness) and its ability to reflect light. The colors of the light reflected by an asteroid, its spectra in infrared, visible, and ultraviolet light, can tell astronomers what its surface is like. Astronomers can pin this down surprisingly well by comparing asteroid spectra with the spectra of meteorites of different kinds. Shiny ones, made mostly of iron and nickel, reflect up to twenty times as much light as the dullest ones. Knowing the brightness of an asteroid and the ability of its surface to reflect light, astronomers can then infer its size. Furthermore, the excellent agreement between meteorite spectra and asteroid spectra provides the strongest evidence that meteorites were once part of larger asteroids.

Sometimes astronomers get lucky. When an asteroid passes in front of (occults) a star, the number of seconds during which the star is invisible depends precisely on the size of the asteroid. If the asteroid's orbit is known, its speed can be calculated, and from that its size can be determined. Astronomers are still awaiting a chance to measure the size of Ceres and most of the other large asteroids this way. But on May 29, 1978, thirty different groups of astronomers watched as Pallas, the second-largest asteroid, occulted a star. They found that Pallas is an ellipsoid rather than a sphere, with a maximum diameter of 559 kilometers.

Eros is the largest of the close-approach asteroids and one of the most intriguing. It is not considered an Earth-crosser (at least, not for now), but at its closest approach, 23 million kilometers, it is visible with binoculars. (Vesta, the fourth-largest asteroid, can occasionally even be seen with the naked eye.) In 1931, observers using a large refracting telescope were able to see the shape of Eros change, apparently as it tumbled. In 1975, Eros passed in front of a star of naked-eye brightness, occulting it for two and a half seconds. From this disappearance and from measurements of its rapid changes in brightness, astronomers concluded that Eros is shaped like a tumbling rounded brick, about 30 by 19 by 7 kilometers in size.

The 1979 movie *Meteor,* showed a cratered "meteor" (we would call it an asteroid) tumbling, much as Eros does, toward the Earth. The collision it depicted is actually quite plausible—the movie was based on an MIT report speculating on the consequences of an asteroid hitting the Earth. But would it be possible for Eros or another *medium*-size asteroid to threaten Earth? A few years back, such a suggestion would have brought from astronomers snickers and scornful comparisons with Velikovsky. Now we cannot be so complacent. Here's why.

The origin of Earth-crossing asteroids has been one of the lingering mysteries of planetary astronomy. For that matter, the origin of the asteroids generally has been controversial. In the late eighteenth century, astronomers first began looking for minor planets in orbit between Mars and Jupiter because of Bode's law, a rule expressing the distances of the planets to the Sun. Now thought of as a mathematical coincidence, Bode's law seemed to suggest a missing planet between Mars and Jupiter. When various asteroids were discovered orbiting at approximately the expected distance, astronomers surmised that the missing planet had broken up or blown apart into these smaller pieces. Later it was found that the combined mass of all the asteroids is far less than the mass of any other planet—which made the surmise of a missing planet much less compelling. In addition, no one could come up with a convincing reason that a planet-sized body would explode.

In the current picture of the early solar system, a primeval solar nebula of dust and gas gave rise to tiny *planetesimals,* or clumps of matter, which joined to one another by gravitational attraction and random collision. In this way most of the large objects in the solar system were formed over aeons of time. But the powerful attraction of Jupiter would have prevented planetesimals from sticking together; most of them would have been either pulled into Jupiter or flung out of the solar system altogether. In certain bands between Mars and Jupiter, however, stable orbits are possible, and it is here that we find most of the present asteroids.

The Earth-crossing asteroids do not quite fit in this picture. Nor do a few dozen other oddballs trapped at two points along the orbit of Jupiter itself, known as the Trojan asteroids. Nor, finally, does a most unusual little world, Chiron, orbiting between Saturn and Uranus. Discovered and named in 1977 by Charles Kowal, Chiron seems to be about the same size as the larger Mars-Jupiter asteroids and may be just one of a whole flock of trans-Saturnian worldlets.

Most asteroids are spherical in shape, for a simple reason. Asteroids are made of rock, and rock is not completely rigid. Given enough pressure, a rocky asteroid deforms. For a planet larger than a few hundred kilometers in diameter, the gravitational attraction of all the pieces of rock to one another is strong enough to pull them all inward and keep them together. Like a liquid drop drawn into a sphere by surface tension, an irregularly shaped large asteroid will eventually become spherical, or nearly so. In both cases a sphere is the most stable state. But many asteroids, like Eros, Ida, and Gaspra, are irregular in shape. Because the inward tug of gravity is less for smaller asteroids, they can retain their ruggedly individualistic shape indefinitely, or until something really big hits them. Astronomers believe that collisions between asteroids gave rise to the fractured, irregular shapes we see, and that some of the minor planets are fragments of titanic collisions between larger bodies.

Where did the Earth crossers come from? At first, astronomers proposed that the mysterious Earth-crossers had been spawned by violent collisions within the asteroid belt. But simple physics raises an objection: When two bodies collide in the absence of outside forces, their center of mass continues moving with the same velocity (according to the law of conservation of momentum). Therefore the center of mass of colliding asteroid-belt bodies must remain in the asteroid belt. For similar reasons, the origin of meteorites is hard to understand, but their spectral resemblance to asteroid-belt material leads astronomers to suppose the belt was their source too.

Some Earth-crossing asteroids may be the remnants of comets that lost their tails and halo, but not most of them. The meteoritic material we find on Earth generally resembles asteroid-stuff much more than comet-stuff, and meteorites, we think, originate from the Earth-crossers.

In recent years astronomers have shown how certain orbits in the asteroid belt, after millions of years of apparent regularity, can suddenly become unstable. The powerful gravitational pull of Jupiter, which influences some asteroids to a degree comparable to the Sun, is an essential element in such destabilization. The laws of chaos apply to these events—the new physics in which small changes in initial conditions lead to big differences in the result, as in the weather. A chaotic system, be it the weather or asteroid orbits, is essentially unpredictable on some level, and not merely because it is complicated. In the orbital dynamics of the asteroid belt, literally millions of gravitating bodies are influencing one another, so astronomers cannot predict which asteroid orbit will suddenly become unstable and catapult a mountain-sized projectile toward Earth. Nor does chaos physics as applied to asteroids violate the law of conservation of momentum, as it might at first appear, since Earth-crossers get their burst of momentum toward Earth from their interaction with Jupiter. (In much the same way, a spacecraft like Galileo picks up enough velocity to reach Jupiter by flying a complex path, acquiring momentum like a slingshot from swinging around an onrushing Venus and twice around Earth.)

The number of unknown Earth-crossing asteroids, as we have seen, far exceeds the 150 or so whose orbits we do know. The Alvarez impact discoveries and Chicxulub have brought a new recognition of the value of finding and tracking small asteroids, especially Earth-crossers. Many planetary astronomers believe that most of the Apollo, Aten, and Amor asteroids will eventually collide with Earth, even if their orbits now do not possess chaotic instability. Nor have we observed them long enough to rule out the possibility of a collision the next time one comes zooming back from its last encounter with Jupiter.

About objects much smaller than a kilometer across—say, 100 meters—we know even less. For the most part, they do not show up in astronomical surveys. A small asteroid on a collision course is even harder to see, since it does not make much of a track—on its final approach it gradually gets brighter. Much beyond Earth-Moon distance (only a few hours away, at asteroid speeds), such objects are virtually invisible to any present optical technology.

CHAPTER SEVEN

Comets

One cloudy night in 1908, when even eager amateur sky-watchers had given up in disgust, a hundred million tons of rock were on a collision course with Earth at 20,000 meters per second.

It was a small comet, less than 100 meters across, with a mass of only a few million tons, about the same as ten supertanker ships. As it streaked across the central Siberian sky on the morning of June 30, few people noticed its fiery trail. Residents of the remote town of Vanavara were startled by a bright flash, then astounded as a huge pillar of fire shot into the sky about 60 kilometers away. Next, a mushroom-shaped cloud billowed into the stratosphere. Were we to see it today, we might think a thermonuclear bomb had exploded, that nuclear war had begun. The people of Vanavara had no such concern, but they felt a blast of intense heat and a thunderous, violent shock wave that broke windows, knocked people from their feet, and collapsed ceilings. Mostly, they were mystified. What had happened, we now know, was the largest extraterrestrial impact of the twentieth-century.

The comet exploded about 8 kilometers above a desolate pine forest in the basin of the Stony Tunguska River—devastating hundreds of square kilometers. As the fiery cloud soared into the stratosphere, it startled peasants up to several hundred kilometers away. Nearly 500 kilometers across Siberia, passengers on a train saw and heard the event. Seismic stations around the world shook. The aerial shock wave circled the globe twice, triggering scientific instruments but causing no damage. In

Europe, thousands of miles to the east, people noticed a peculiar glow in the night sky. Soot from the fire blew across the Pacific Ocean and darkened California skies.

Directly under the blast, trees lost their limbs but stood. Off center, the blast knocked trees down in a symmetrical pattern pointing outward from ground zero. Up to 20 kilometers away, most trees were flattened; some were knocked down as far as 40 kilometers away. Strangely, the trees were scorched but not completely burned. Today some scientists believe that intense heat from the initial explosion set the trees ablaze but that the blast wave blew the fire out. One farmer was thrown off his porch by the blast, but no one was seriously hurt.

Based on the extent of destruction to the forest and the damage in Vanavara, it is possible to estimate the energy of the blast—about the same as a 10-megaton thermonuclear bomb. Had ground zero been the center of a major city instead of a lonely forest, the downtown area would have been flattened and set afire; casualties might have reached the millions. Even people in the surrounding countryside would have been knocked cold by the blast wave.

Relatively few meteoritic fragments were ever found in the Tunguska forest. No crater was formed, which is why we think the blast took place high in the air. Expeditions eventually found small quantities of glassy and metallic spherules, which could have condensed from vaporized comet dust. The bright night sky seen from afar may have been caused by this dust lingering in the atmosphere. Compared with famous comets that had spectacular glowing tails, the Tunguska comet was puny. Actually, some scientists think an ordinary stony meteorite about 80 meters in diameter is more likely the cause of the Tunguska explosion. The evidence is not decisive either way. Nor is the distinction between comets and meteoroids and asteroids absolute—old comets may be indistinguishable from asteroids. Whether it was an asteroid or a comet, an object the size of the one that devastated Tunguska, coming right at us, it would be difficult to detect telescopically. The next one might might strike anywhere on Earth with absolutely no warning.

Throughout recorded history, comets have been swathed in superstition. Although beautiful, they were often considered flaming omens of doom—portents of famine, disease, revolution, or defeat in war. About once a decade a comet bright enough to be seen with the naked eye appears. Typically a comet looks like a fuzzy patch in the sky, with a bright head and a long tail that points away from the sun. They can remain visible in the night sky for many weeks. About once in a century there is one so bright, it can be seen in daylight. Comets do not appear to be speeding across the sky like meteors. Rather, they seem to hang oddly suspended in place. Viewed from night to night, they can be seen to have moved slightly with respect to the stars. This movement distinguishes them from galaxies and nebulae, which appear as faint but unvarying smudges in the night sky. Astronomers, both professional and amateur, discover and track them at a rate of about a dozen per year.

Most comets are in orbit around the sun on very elongated elliptical paths. They spend nearly all of their multimillion-year orbital periods in the distant reaches of the solar system, many thousands of times farther away from us than the Sun. As a comet approaches the Sun, solar heat vaporizes the frozen gases it contains; its tail enlarges enormously as radiation pressure forces the glowing gases backward. The spectacular swept-back tail we see is thus the merest trace of gas, spread out over a vast expanse of space. Only the cometary nucleus is solid, composed largely of frozen water, ammonia, carbon dioxide, and methane ices, along with some more exotic (and poisonous) ices such as frozen formaldehyde and cyanides. Mixed in with the ices are dust and rock, totaling perhaps one-third the total mass. Harvard comet expert Fred Whipple, who first proposed this structure, called comets "dirty snowballs."

As a cometary snowball, or perhaps more appropriately rocky iceberg, rounds the Sun, its tail of glowing ionized gas and dust continues to stream away from it. Seen through even the most powerful telescopes, the solid core is but a tiny pinpoint of light. Until 1986, when an armada of spacecraft observed Halley's comet at close range, no cometary nucleus had ever been resolved visually. Even when Halley's nucleus passed directly in front of the Sun in 1910, it was too small to be visible. Its position is more or less obvious, however, for as the nucleus heats up while hurtling toward the Sun, a huge luminous sphere of gases forms around it. This head or *coma* can be a million kilometers or more in diameter, dwarfing the nucleus. Nuclei larger than 20 kilometers across are probably rare, and most of them are much smaller. Surrounding the coma but invisible to the eye is the hydrogen envelope, which glows mainly in ultraviolet light.

Cometary tails come in a wonderful variety of sizes and shapes. Some comets have short thick tails; others have wispy thin ones that could span the entire 150-million-kilometer distance from Earth to Sun. The tail may be broken by multiple streaks and complex splits. Bright fountains and jets of gas squirt periodically out of the head and mix with the tail. Usually comets have two tails, one blue and predominantly made of ions (atoms that have lost electrons) and the other mostly yellowish dust. The ion tail tends to be straighter because its constituents move so fast, and its appearance is variable from one night to the next. The dust tail, composed of slower particles, arcs away from the sun and spreads out more. Sometimes it looks like a multilayered dome or shroud. Ancient observers carefully categorized the different shapes of comet tails and confidently associated them with specific evils. Almost always such prognostications were wrong, but in that prescientific era, it was not considered necessary to check whether predictions came true, let alone modify the belief system that had generated them.

The possibility that a comet could collide with Earth and bring doom did concern some historical observers. In the nineteenth century astronomers also realized that the Earth might pass inside a comet's tail. They had detected the presence of organic molecules in cometary gases, even potentially poisonous molecules such as cyanogen. Just before the arrival of Halley's comet in 1910, a huge wave of fear swept Europe and North America, as people imagined themselves suffocating or dying a horrible death from cyanide poisoning. But the hysteria was groundless. The density of matter in a comet's tail is extremely low; the cyanogen and other exotic molecules there are far too sparse to pose a threat.

A fraction of the comets discovered each year are short-period comets, like the famous Halley's comet; they circle the Sun in three to two hundred years. One of the shortest-period visitors is Encke, with a return time of three years and four months. Astronomers have been tracking Encke for nearly 150 years; they found long ago that its elliptical path lies entirely within the orbit of Jupiter. Circumstantial evidence points to a fragment from Encke as a possible source of the devastation in the Tunguskan basin in 1908. In that year Encke's orbit changed, as if a chunk had broken off.

As comets go, Encke is just a baby. Radar probes have measured its rotating solid core to be no more than 2 kilometers across, about the size of a large iceberg or a mountain peak like the Matterhorn. Once trapped in the inner solar system, a comet like Encke will live no more than a few thousand years. The sun's radiation will completely vaporize its ices, leaving only dust and rocky meteors. If the comet strays too close to Jupiter, the giant planet will capture it, leading possibly to a collision, as in the case of comet Shoemaker-Levy 9.

The brilliant astronomer Edmund Halley was the first to show that comets can return. In a remarkable mathematical feat for his precomputer, precalculator age, he analyzed the orbits of twenty-four comets recorded between 1337 and 1698. Using Newton's newly discovered laws of motion, he was able to demonstrate that the spectacular visitors of 1531, 1607, and 1682 (the latter he had seen himself) were one and the same object, a comet with a period of seventy-six years that was later to bear his name. He could even account for small differences in the orbit of the comet that were observed in those three appearances by estimating the gravitational influences of Jupiter and Saturn. Having measured the period of this bright comet, Halley predicted its return in 1758 (and 1834, 1910, 1986, and so on). Scholars have subsequently been able to find records in one country or another matching every approach of Halley's comet since 239 B.C.

On Christmas night in 1758, a German amateur astronomer named Johann Palitzsch was the first to witness the comet's predicted return. It was an enormous triumph not only for Halley but for Newton. The comet's latest return, in 1986, was a different sort of triumph: No fewer than five spacecraft flew near the comet to take close-up pictures and gather data. The European Space Agency's Giotto spacecraft recorded images of Halley's comet from only a few hundred kilometers away. They show a dark and irregular-shaped nucleus about 8 by 15 kilometers in size, roughly the dimensions of San Francisco. Interestingly, this is also about the right size required to blast out the Chicxulub crater.

Some comets probably have much larger nuclei, which at several hundred kilometers across approach the size of major asteroids. One of the brightest comets ever recorded was the Great Comet of 1729, easily visible to the unaided eye. Its point of closest approach to the Sun (perihelion) was actually unusually far away, almost at the far extreme of the asteroid belt. Only a very large object could have appeared so bright when so far away.

Other comets approach so close to the Sun that they almost hit it. In 1965 comet Ikeya-Seki came within 1.2 million kilometers of the Sun. This does not sound so close—until you realize that the Sun is only about 1.4 million kilometers across. So strong are solar tidal forces at such a distance that Ikeya-Seki was torn into two fragments. The Great Comet of 1680 came even closer, to within 100,000 kilometers, but curiously, it did not break apart. Comet Howard-Koomin-Michels, discovered in 1979, came too close: After rounding the Sun, it emerged headless, while its tail remained visible for several days before breaking apart and disappearing.

Presumably some comets make direct hits on the Sun, but such an event has yet to be observed. Even if a comet doesn't break up or collide with anything, each pass around the Sun vaporizes more of its ice, revealing deeper and perhaps older layers of frozen materials. At the same time, more comet dust is blown off into space. Comets are "mortal" in this sense: after a certain number of returns, they will be reduced to mere rocks, unable to form beautiful tails. At this point they become indistinguishable from asteroids, except possibly from their orbits. Several of the Earth-approaching asteroids do have orbits that resemble those of known short-period comets.

Even when no comet is near Earth, the effects of comets are visible. Comets have filled the solar system with dust. Light scattered from space dust is visible after sunset, although not to city dwellers. You can see this "zodiacal" light only on very dark nights, in locations far away from cities and their bright lights. Look for a faint glow above the horizon, near the position where the Sun set. On the horizon opposite the sunset, you may also be able to see a glow from the same source; this light is known as the Gegenschein, or opposite glow. Both the faint glow and its Gegenschein come from the scattering of sunlight by dust, evidently left over from the passage and breakup of comets. Some of the dust may have been blasted into space by the powerful asteroid impacts that formed craters on planets and their moons. Some of this rebounding material acquires escape velocity and does not return to its planet of origin. This solar system dust must constantly be replenished because sunlight slows it down (a discovery of physics called the Poynting-Robertson effect) and it eventually spirals into the sun; that the glow persists means that more dust is being generated.

About a thousand comets have been seen by astronomers or other observers in recorded history, most as faint smudges on a single pass around the Sun. Short-period comets, with recurring appearances, are actually in the minority. Long-period comets have orbits so elongated that they spend most of their lives at a very great distance from Earth. Based on the rate of observation—about half a dozen per year—and other data, it is possible to estimate the total comet population, which is now believed to be in the trillions. For hundreds of years, astronomers speculated that vast numbers of comets were undiscovered. Measurements of comet orbits revealed that short-period comets orbit within the same plane as the planets and their moons (known as the plane of the ecliptic). Most travel around the Sun in the same direction as the planets—comet Shoemaker-Levy 9 was a conspicuous exception to this rule. By contrast, the orbits of long-period comets can lie in any plane, so that they occupy a spherical region of space around the Sun so huge, it extends more than halfway to the nearest stars. They are as likely to orbit the Sun in a direction opposite that of the planets as in the same direction.

But up until this century, scientists did not understand what launched comets from extremely distant regions on trajectories that brought them close to the Sun.

During the 1950s Dutch astronomer Jan Oort showed that the gravitational effect of comets' repeated interactions with nearby stars scrambles the comets' orbital planes into a nearly random mixture. This scrambling is why long-period comets approach the Sun from all directions. Our Sun, itself a star, moves within a region of space inhabited by many other stars. It carries with it on its journey through our galaxy the planets and the lesser members of the solar system, including comets. All are gravitationally bound to the Sun and have little chance to escape it. But each time the Sun approaches another star, the comets receive a little push, or perturbation. Perturbations due to successive star encounters will eventually elongate the orbits of an "unlucky" few, out of a vast number of comets, and trigger their fall toward the Sun. As a result of being "kicked" by passing stars, other comets will be ejected from the solar system altogether.

Comets awaiting such a series of fateful perturbations orbit the Sun in cold storage, in a region now called the Oort cloud. Denizens of the cloud, although called comets, are dormant, solidly frozen bodies, without a hint of the glories possible in the glare of the Sun. Current theory holds that there are at least three regions within the comet cloud. The outer Oort cloud, containing about 10^{12} comets, fills a volume of space of about 20,000 to 50,000 Earth-Sun distances. (The Earth-Sun distance, equal to 150 million kilometers, is called 1 astronomical unit, or AU.) In other words, the outer Oort cloud comprises an extremely distant region of the solar system. Perhaps ten times as many comets are trapped in an inner Oort cloud, stretching from 20,000 AU inward to about 3,000 AU. Between 3,000 and 100 AU there are few comets. Inward of this, however, is another comet haven called the Kuiper cloud. Recently, using the space telescope, astronomers have observed "sleeping" comets in the Kuiper cloud. Close-up study poses severe problems—even if a spacecraft were to be launched successfully into the comet cloud, astronomers would have to wait thousands of years for the data to be radioed back to them. Once the spacecraft reached the cloud, it would be difficult to get it to locate comets, because the comets are so far apart.

It's tempting to think of the Oort cloud as packed with comets, with only the rare comet to be encountered passing across the inner solar system. Actually, it's just the other way around. So vast are the Oort clouds that the average distance between comets there is many times the size of the inner solar system with its planets. On the other hand, at any one time there are hundreds of comets in the inner solar system, most belonging to the short-period family; only one or two are genuine long-period comets. With this insight in mind, the Oort cloud becomes a staggeringly lonely place, cold beyond imagining—average temperatures are only a few degrees above absolute zero. From the icy surface of one slowly tumbling comet, it is impossible to see another. The sky is always black, lacks planets or moons, and contains exactly one more dim star than our night sky, the Sun.

Most planetary astronomers think the short-period comets like Encke and Halley were once trapped in the Kuiper cloud. Just as the stars kick

Oort cloud comets toward the Sun, so the gravitational influence of Neptune launches Kuiper comets into near Sun-crossing orbits, along which they may also be deflected by Jupiter into even smaller orbits like Encke's.

While people do have reason to fear comets, they have value for us as well, and not only for their beauty. Comets present us with a precious gift; a sample of matter largely unchanged since the birth of the solar system 5 billion years ago. Comets are generally regarded as made up of material left over from the formation of the solar system 4.5 billion years ago. They are thought to have been formed by the gravitational collapse of dust and gas clouds, as were the planets. Whereas most bodies in the comet size-range (about 1 kilometer across) collided and merged to become part of planets, some in highly eccentric orbits met another fate. Chance gravitational encounters with the young giants Jupiter and Saturn kicked them out of the inner solar system, and thus were the Kuiper and Oort comet clouds born.

In all, the 10^{13} (10 trillion) suspected comets, each averaging one or a few kilometers in diameter, may have a combined mass only a few tens of times that of the Earth. Although their numbers are impressive, their small total mass makes them relatively minor constituents of the solar system. (Jupiter alone is 318 times the mass of the Earth, and the Sun is a thousand times more massive than Jupiter.) The combined mass of comets (as well as their total number) is far greater than that of asteroids—at least today's asteroids: All the asteroids together total less than one percent of the mass of Earth. On the other hand, asteroids may seem the greater menace to us because they orbit closer to Earth than most comets. We still do not know enough to be sure whether a deadly comet strike or an asteroid collision is more likely on Earth today. Most of the monster craters on other bodies in the solar system—larger than any now known on Earth—were probably made by ancient rocky asteroids rather than by icy comets. These craters are very old. Some very large lunar craters have been dated, using rocks brought back by the Apollo astronauts, to about 4.5 billion years old on the average. Collisions of world-shattering force were more common in the chaotic early history of the solar system, more than 4 billion years ago, when large rogue asteroids orbited the sun in wildly eccentric orbits. The bombardment continues at present, but the largest and most deadly missiles have already hit something or been ejected from the solar system. Most of the smaller ones have long since been swept out of the solar system as well.

This information may be reassuring, but a collective sigh of relief would be premature. The mass extinction of 65 million years ago, one of the two or three most deadly in the fossil record, is actually a relatively *recent* event. There has been little evolution of the solar system since then. Of the larger missiles, few have been eliminated in the last 65 million years, compared with the major housecleaning that took place in the preceding 4.4 billion years. Smaller comets and Earth-crossing asteroids are replenished at about the rate that they collide with planets or get ejected from the solar system. So the odds of a deadly collision cannot have changed much since the dinosaurs vanished. The odds of a mass-extinction-causing-catastrophe that blasts a 200-kilometer crater are still about one per hundred million years. Lesser catastrophes leaving only a 100-kilometer crater may occur *on average* about every 20 million years. What we are still unsure of is how the collisions are spaced. For example, we do not know whether comet crashes come in storms, or if major impacts are random, unpredictable events. One body of evidence, based on the fossil record, suggests that catastrophic impacts are far from random, that they happen as part of a deadly pattern.

CHAPTER EIGHT

Nemesis and Mass Extinctions

At crucial times in the fossil record, large numbers of apparently thriving species became extinct, all within a relatively short interval. Different creatures replaced them, living in similar climatic conditions, in the same geographical areas. These successor species were not necessarily better adapted or more "fit"—they just appeared later in evolutionary history. Of the many extinctions known, five stand out as more globally devastating than the others. Marking the boundaries of epochs in Earth's geological and biological history, these five great extinctions took place about 440, 365, 245, 210, and 65 million years ago. Of these, three extinctions have been linked with large Earth craters and three with boundary clays rich in iridium, suggesting an extraterrestial impact.

The last great extinction, the K-T catastrophe 65 million years ago, saw the end of about 40 percent of all animal genera (*genera* is the plural for *genus*, the category in biologists' classification system between family and species) and about two-thirds of animal species. Marine reptiles, including the huge long-necked plesiosaurs, the mosasaurs with paddles instead of legs, and the sharklike icthyosaurs, died out completely. Every single species of terrestrial dinosaur vanished; the birds, considered by many biologists to be a form of dinosaur, survived. Most flowering plant species endured and made a comeback after ferns took over the continents. (Such a takeover is exactly what happens when a forest burns today: Eventually young trees crowd out ferns, which cling to existence below the forest canopy.)

The fossil record does not tell clearly whether these great extinctions happened in a single day or over several million years. Geology is too uncertain a science for that, at least for now. Erosion makes the record difficult to read. Dating methods are too imprecise to allow paleontologists to compare reliably the fossil record at different sites around the world. Fossils of large animals like dinosaurs are rare. At one key site a 2- or 3-meter gap lies between the most recent dinosaur skeleton and the iridium-rich clay layer. As Luis Alvarez and many others have pointed out, this gap does not necessarily mean that the dinosaurs died out before the impact. There might simply have been an interval during which no dinosaur was preserved well enough to become a fossil in the rather small fraction of the Earth's surface that has been searched. Since the average spacing between dinosaur skeletons is only about a meter, this possibility is statistically reasonable. In any event, we have no reason to expect a concentration of dinosaur fossils right at the K-T boundary.

In between the five great extinctions were many others—almost two dozen, according to paleontologist David Raup. In these lesser events, smaller percentages of species and genera became extinct. Two smaller extinctions, the ones dating from 35 million and 290 million years ago, have been tied to known craters.

Before the Alvarez work linked the K-T catastrophe with an impact, paleontologists had been debating the causes of mass extinctions for decades. Despite the general feeling that no one mechanism could possibly dominate in such a complicated process, climatic change, especially cooling and drying, was the most popular explanation. There have been many others, including the rises and falls of sea level, predation, epidemics, competition with other species, poisoning of ocean waters, changes in atmospheric chemistry, global volcanism, and comet or asteroid impact. Established species that are well distributed geographically are extremely difficult to kill off, Raup emphasizes in his studies of extinction; he came to the conclusion that some kind of extraordinary "first strike" was necessary before most of the proposed extinction mechanisms could have a credible chance to work. Was it possible that extraterrestrial impact has been the *major* cause of mass extinctions, maybe even the only cause?

University of Chicago paleontologist John Sepkoski had a special interest in the dates when particular types of fossil appear in the record and when they disappear. In 1982, after collecting fossil data for many years, he produced a massive compendium of 3,500 families and by 1984 a computerized index of 30,000 genera. Raup and Sepkoski suspected that this mass of data contained some simple patterns that could shed light on the mechanism of mass extinctions, but they had no particular mechanism in mind at that time. In 1977 paleontologist Al Fisher had claimed to find a regular 32-million-year spacing between mass extinctions. Using a variety of computer methods to analyze the twelve biggest mass extinctions, Raup and Sepkowski found a regular spacing too, but at 26 million years rather than 32 million. Try as they might (and as good skeptical scientists must), they could not shake the regularity of the intervals from their calculations.

A dozen or more groups have reanalyzed the Raup and Sepkowski data to see if mass extinctions do have periodicity. According to David Raup, "The results are mixed: about half support the 26-million-year periodicity (with minor revisions of the period in some cases), and about half find no convincing evidence for cycles of any duration." Raup himself still believes the periodicity is real, but most paleontologists do not. Another, more subtle objection is that the periodicity in the fossil data is real but is the result of a characteristic "healing time" that life needs in order to recover after any deadly impact rather than a regular period of the impacts themselves.

The clocklike spacing between extinctions was nonetheless intriguing. One of the present authors, Richard Muller, caught wind of the Raup and Sepkowski finding before its publication. Muller came up with a possible explanation—our Sun may have had a little companion star orbiting around it with a 26-million-year-period. *Most* stars, after all, are part of binary systems. The two stars closest to Earth, Alpha Centauri and Proxima Centauri, orbit each other. If our Sun's hypothetical companion star approached the inner solar system every 26 million years, it might knock lots of asteroids out of their usual orbits. One or more of these asteroids could hit the Earth and cause extinctions. Not only did this explanation provide a plausible clock in the sky, it had an important side benefit: The asteroids could come in bunches. This would answer paleontologists who kept objecting to the impact theory on the grounds that the dinosaurs had died off over hundreds of thousands if not millions of years, rather than all at once. Okay, maybe it took several crashes to do in the dinosaurs, astronomers would be able to reply.

Unfortunately, this first explanation of the periodicity of mass extinctions had a fatal flaw: An orbit that brought a companion star close enough to the Sun to enable it to fling asteroids from orbit would be very elongated and unstable. The pull of passing stars would change the orbit so much that on the next pass it would come nowhere near the inner solar system. A changing orbit could not explain periodicity.

Muller, teaming up with astronomers Marc Davis and Piet Hut, soon came up with a workable revision of the theory. Suppose the companion star were on a much less elongated orbit, one shaped roughly like an egg. Its maximum distance from the Sun would be about 3 light-years and its minimum distance about half a light-year. (Half a light-year may not sound like much, but it is a distance 160 times greater than the orbital distance of Pluto from the Sun.) This more circular orbit would be much more stable and could produce periodic impacts.

Every 26 million years, the companion would pass through the Oort comet clouds. There, just as Oort had theorized for randomly passing stars, it would destabilize the orbits of billions of comets. Some would gain energy and speed and be ejected from the solar system. Others would lose energy and begin a long fall toward the Sun. For every billion comets dislodged, the team calculated, about a million would cross Earth's orbit. Of these, about two would hit the Earth. The numbers came out right: The bombardment would last about a million years, and during that time a new comet would be visible from Earth about every three days. Very few of these would hit. For each complete cycle, sometimes one comet would hit Earth, and sometimes two, three, four or five. Sometimes, by pure chance, none would hit.

Muller suggested that the companion star be named Nemesis, after the Greek goddess who made sure no earthly beings challenged the dominance of the gods. Before the startling new hypothesis could be published, however, a serious question had to be addressed: Would the

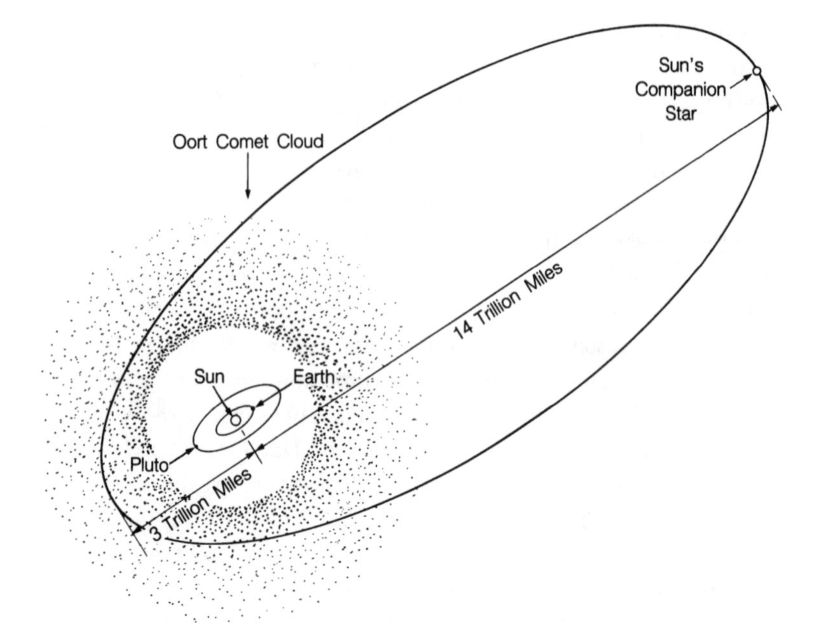

Figure 8-1: Hypothetical orbit of Nemesis, companion star to the Sun, showing how it passes through the Oort comet cloud every 26 million years. If it exists, Nemesis would currently be near apogee, about 14 trillion miles from the Sun.

companion star's orbit be stable, or would it too be disrupted by passing larger stars? Piet Hut published a calculation that showed that the duration of the present Nemesis orbit (its "lifetime"), if it existed, would be about 1 billion years. This means that in the next billion years, there's about a fifty-fifty chance of a passing star kicking Nemesis out of its liaison with the Sun, thus ending its reign of terror. The billion-year figure is based on the *present* size of the orbit of Nemesis. According to Hut's calculation, the orbit of Nemesis has grown steadily over the 5-billion-year lifetime of the solar system. The companion star was once much closer in, with a correspondingly shorter orbital period. When Nemesis was originally formed along with the Sun and planets, its orbit *then* lasted about 5 billion years. Encounters with passing stars, on average, have tended to increase the energy of Nemesis so that its orbit lengthens. This is analogous to what Nemesis does to comets in the Oort cloud: It boosts their energy so that more of them leave the solar system than the number that lose energy and fall inward.

If the Nemesis theory is correct, geologist Walter Alvarez quickly realized, there should be evidence for it in the cratering record on Earth, such as a regular pattern in the impact dates. He and Muller began searching for periodicity in the ages of the impact craters on Earth. The first plots they studied were disappointing: No pattern jumped out at them. But many craters are very poorly dated. Uncertainty in their ages could easily blur a 26-million-year period. Alvarez's solution was simple: Ignore the craters that don't have accurate dates. When the list of nearly a hundred craters was culled down to two dozen, something exciting happened. There were three or four obvious pairs spaced by 30 million years or so, and these included some of the largest craters. When only the larger craters were plotted, clumps of them seemed to be spaced by 26 to 30 million years. The average interval was 28 million years. A detailed statistical analysis came next. Fourier analysis, a powerful mathematical technique that finds periodicity in data, produced a strong peak at a period of 28.4 million years, towering above any other possible regularity in the data. When the computer was called upon to calculate crater ages distributed randomly, the Fourier program found a comparably large peak only every few hundred tries. That is enough statistical significance to be interesting, but it is not overwhelming evidence for a new theory.

Claimed "discoveries" in science have a long history of appearing to be moderately convincing on a statistical basis but going away later when more data come in. In the long run, it doesn't help a scientist's reputation to be involved in making such a claim, even if the discovery paper catapults him or her to overnight fame. In this case, Walter Alvarez and Rich Muller were well enough known already that they probably had more to lose than to gain from publishing the Nemesis theory. Luis Alvarez, for his part, had seen plenty of these scenarios, and he certainly wanted to protect his son and Muller from a possible mistake. He had been alternating between extreme enthusiasm for and skepticism about the Nemesis theory, but finally he tried vigorously to discredit the periodicity in the crater data by showing that the data were not statistically significant or were otherwise flawed. After several intense weeks of give and take with his son and Muller, Luie relented, and his skepticism abated. The Nemesis paper was sent to the journal *Nature* for publication.

Louisiana astronomers Daniel Whitmire and Albert Jackson had independently come up with a similar theory to explain the apparent periodicity of mass extinctions—comet showers triggered by a companion star—but they hypothesized an extremely eccentric (and, it turned out, unstable) orbit. They sent their paper to *Nature* too. Whitmire also advanced the idea of a Planet X orbiting beyond Pluto. Such a planet, he showed, could perturb the inner part of the comet cloud and cause comet showers. But Planet X would spread the comet showers over many millions of years, it turned out, in contradiction to the mass extinction data. Another theory focused on the bobbing motion of the Sun in and out of the galactic plane. Well known to have a period of about 33 million years, this motion could cause periodic perturbations of comets in the Oort cloud, due to the concentration of stars in the plane and the lack of stars out of it. Unfortunately for this theory, there is no correlation between the actual crossings of the galactic plane by the Sun and the ages of the mass extinctions. Also, the bobbing of the sun is too small to give the necessary effect.

One could hardly imagine a more sensational and controversial scientific hypothesis than an invisible killer star orbiting the Sun flinging deadly missiles at Earth. Scientific journals in 1984 were filled with earnest debate, but some of the utterances of usually polite scientists were, to put it diplomatically, less than temperate. The explosion of media interest was incredible. *Time* magazine put the story on its cover. There were TV documentaries, endless TV interviews with the scientists involved, even editorials in *The New York Times.* In 1985, one of these, entitled "Miscasting the Dinosaur's Horoscope," concluded

> *Terrestrial events, like volcanic activity or changes in climate or sea level, are the most immediate possible causes of mass extinctions. Astronomers should leave to astrologers the task of seeking the cause of earthly events in the stars.*

Like the tides and seasons? Walter Alvarez and Richard Muller replied in a letter to the *Times.*

> *You say, "Complex events seldom have simple explanations." The whole history of physics contradicts you. You suggest "Astronomers should leave to astrologers the task of seeking the causes of earthly events in the stars." May we*

suggest it might be best if editors left to scientists the task of adjudicating scientific questions.

The famous biologist Stephen J. Gould lampooned the *Times* with this piece of parody, supposedly taken from an Italian newspaper of 1663:

Now that Signor Galileo, albeit under slight inducement, has renounced his heretical belief in the earth's motion, perhaps students of physics will return to the practical problems of armaments and navigation and leave the solution of cosmological problems to those learned in the infallible sacred texts.

Carl Sagan found the Nemesis theory

serious and respectable, if highly speculative, science, because the principal idea is testable: You find the star and examine its orbital properties.

Sagan wrote his own letter to *The New York Times* defending the Nemesis theory. For several years the debate raged inconclusively.

Today, the concept that extraterrestrial impacts cause devastating catastrophes on Earth is well accepted. Also extremely persuasive is the evidence linking the K-T mass extinction with the impact at Chicxulub crater.

Many people, including astronomers who should know better, believe that the Nemesis theory has been disproved, that the orbit has been shown to be unstable. Of course the orbit *is* unstable, with an expected lifetime of about a billion years, as shown in the orginal Nemesis paper and as subsequently verified in detail by Hut. Since the solar system is 5 billion years old, many people think that a billion-year orbit could not still be in existence; the orbit would have lasted only 1 billion years into the life of the solar system. But they are confusing the *present* lifetime of the orbit with its past duration. The lifetime of the Nemesis orbit has been slowly decreasing since the formation of the solar system. Five billion years ago, Piet Hut showed, the expected duration of the Nemesis orbit was 6 billion years. According to the theory, Nemesis has only 1 billion of these years left.

There may be good reason to question the Nemesis theory, but the obvious question is, Why hasn't Nemesis itself ever been seen? At a distance of three light-years from us, it would be the closest star, more than a light-year closer than the Centauri pair. Basically the answer is that

Nemesis, if it exists, is too small and too dim. (We are assuming that Nemesis is an ordinary red dwarf, like most visible stars much smaller than the Sun.) In order to have enough gravitational kick to launch comets at Earth, Nemesis would have to have a mass of at least one-twentieth that of the Sun. If its mass were as large as one-third that of the Sun, it *would* have been seen—in fact, it would be brighter than Proxima Centauri, and its extreme nearness to us would be well known. But if its mass were much less than a third of solar mass and its brightness correspondingly reduced, Nemesis would be an obscure dim star that looked just like many intrinsically brighter stars much farther away. Its proximity would probably have been missed by standard astronomical surveys. Until the Nemesis theory, astronomers had little reason to make the measurements necessary to reveal the closeness of a dim star.

To convince the world (and ourselves) that Nemesis is real would require discovery of the star itself. Hunting for Nemesis, however, is like looking for the proverbial needle in a haystack. Astronomers measure the distance to nearby stars using a method based on the phenomenon of *parallax.* To grasp this concept, extend your finger in front of you and close one eye. Note the position of your finger relative to some object in the background, such as a picture on the wall. Then switch eyes. Your finger seems to jump. It has a different position with respect to the fixed background as seen by your other eye. That's parallax. Astronomers observe a star at three-month or six-month intervals and carefully measure its position relative to other stars, especially ones known to be very far away. What they are doing is observing the star from different parts of Earth's orbit around the Sun. The position of a nearby star is measurably different when seen from different sides of Earth's orbit. Finding Nemesis requires surveying thousands of stars at different times of the year, then comparing the images with great precision. Using an automated telescope to survey the northern hemisphere, the Berkeley group has eliminated more than half of 3,100 candidate red dwarf stars. On every clear night about ten stars can be checked off. The search will go on until no candidates are left—or Nemesis is found.

Nemesis may not be a red dwarf at all. It could be an exotic object like a black hole, a neutron star, or a brown dwarf. Gravitationally, its action in dislodging comets while circling the Sun would be the same as a red dwarf's. But it would be next to impossible to detect, and we have no reason to suppose that such strange objects are common in this part of the Milky Way galaxy.

Nemesis may not exist at all, and impacts may not even be periodic. Could mass extinctions still be caused by comets and asteroids? Paleontologist Raup thinks they could. As we mentioned earlier, he thinks a "first strike" must reduce the geographical range of a healthy species before it can become vulnerable to extinction. Comet showers require neither Nemesis nor periodicity. Any passing star that perturbed the orbits of lots of comets could produce a deadly shower.

Nor is the geological record clear on the link between impacts and extinctions. Of seven boundary layers in the Earth where iridium is in excess, Raup finds that four are associated with important mass extinctions, but the others are questionable. And of fourteen craters at least 32 kilometers in diameter and younger than 500 million years, five (including Chicxulub) match the ages of mass extinctions. This information is intriguing but inconclusive. The greatest mass extinction of all, the one 245 million years ago, has not been linked to any impact. How any major impact could fail to produce extinctions is unclear, given the enormous release of energy and long list of atmospheric horrors that accompany it. But what is a "major" impact? We don't know the impact energy threshold above which mass extinction is inevitable—if we tried to guess, we could be wrong by a factor of ten, even a hundred. There are other variables, too, that affects an impact: the type of rock struck helps determine the nature of the dust cloud and the intensity of the killing acid rain that accompanies it. The very largest known crater, Chicxulub (170 kilometers), is definitely associated with mass extinctions, and so is the second largest, Manicouagan (100 kilometers) in Quebec, which is linked to the 208-million-year-old great extinction between the Triassic and Jurassic eras. The craters not yet matched with extinctions are in the 50-kilometer range; the energy needed to make them is at least ten times less than that required for Chicxulub.

If the Nemesis theory *is* correct, where do we stand now in the extinction cycle? The most recent mass extinction occurred about 14 million years ago. If Nemesis were to blame, it must have made a pass through the Oort cloud about 14 million years ago. Now it would be near its farthest distance from the Sun, destined to return to the cloud in about 12 million years. So for now we are safe—at least from Nemesis. But the Nemesis theory does not claim that *all* large impacts are due to comets dislodged by Nemesis, or even due to comets at all. Some may be caused by rogue asteroids. Don't get too comfortable! For although we humans owe our "successful" evolution to an impact, we might someday owe our doom to one too. What can we do to guard against this frightening possibility?

CHAPTER NINE

Spaceguard

After the 1980 Alvarez proposal that an asteroid impact caused the K-T mass extinction, NASA became interested in detecting near-Earth asteroids and deflecting them from a collision course. Using existing telescopes of modest size, astronomers intensified their efforts to find Earth-crossing comets and asteroids—with great success. Since 1980, the number of known Earth-crossing asteroids has more than doubled to over 150. The discovery rate is accelerating. During the late 1980s, asteroid hunters detected several near misses by mountain-sized objects. In 1990, Congress authorized more NASA studies. Thanks to the involvement of dozens of scientists from several nations, an increasingly refined picture of impact hazards is emerging.

Obviously, the bigger and faster a projectile, the more dangerous it is. Meteorites are usually no larger than a person's fist. By the time they reach the surface of Earth, these chunks of rock or iron have been much slowed from their velocity in the solar system—only those made mostly of iron strike with a large fraction of their entry speed. Very rarely they crash into houses, and in a few instances they have injured people. Bodies in the 1-to-10 meter size range rarely reach the ground without fragmenting. In 1972, a small asteroid about 10 meters in diameter left a flaming trail 1,500 kilometers long over the American West. Even such an asteroid carries the kinetic energy of a Hiroshima-sized atomic bomb, equivalent to about 13,000 tons of TNT explosive. An iron body of similar size blasted a small crater in Siberia in 1947. Since most of the Earth's surface is sparsely inhabited or open ocean, the majority of these kiloton bursts go unobserved.

Far more dangerous are cometary or asteroidal bodies in the 50-to-100-meter size range, like the one that exploded over Tunguska in 1908. With kinetic energy in the multimegaton range, a Tunguska-sized airburst could level a large city and kill many people. But the damage from a Tunguska-style event would be purely local; people 500 kilometers away might see it, but in no way would they be threatened. We can expect a 10-megaton airburst somewhere on Earth about once every three hundred years; and on average every hundred thousand years (given present population densities), one will devastate a heavily populated area

Asteroids in the next-largest size range, up to a kilometer across, have a good chance to penetrate the atmosphere without fragmenting; those of more than 150 meters in diameter hit about once every five thousand years, and if they strike land, they almost always create a crater over 2 kilometers across. Those hitting in the ocean cause tsunami waves. Events in this size range are less destructive than Tunguska-style airbursts since most of the impact energy is absorbed in the ground or ocean. Nevertheless, an asteroid almost a kilometer across could decimate hundreds of thousands of square kilometers. If ground zero were in an urban area such as New York, southern California, Tokyo, or greater London, the death toll could exceed 10 million. Still, humanity as a whole would not be threatened.

The effects of an asteroid or comet bigger than a kilometer across, which hit land about once every half a million years, would be global. Dust thrown up by such an impactor might threaten starvation for much of the world's population, due to massive crop failures. To what extent institutions and nations would survive such a global catastrophe, no one can say.

And yet no matter how awful the prospect of a civilization-threatening global catastrophe, such an event would extinguish few species of life. Still, once every 10 to 30 million years an asteroid or comet with a diameter of over 5 kilometers arrives on target Earth. And once every hundred million years, we suffer collision with an object in the 10-kilometer-and-up-range. From the impacts of such large bodies, the planet experiences not just extinctions but *great* mass extinctions, like the "big five" well known to paleontologists. Probably a collision with the largest known

Earth-approaching objects is the ultimate horror. Halley's comet, as we have seen, has a nucleus about 15 kilometers in its longest direction. The largest *known* Earth-crossing asteroid is a little smaller than that but is probably several times more dense, and therefore more massive. But we can't rule out the sudden appearance of a long-period comet somewhat bigger than Halley's, on collision course with Earth.

It is tempting to discount the risk of all the above disaster scenarios, simply because we have no record of such a deadly impact. No victims of an asteroid strike appear on TV to arouse our sympathy. The hurricanes, floods, earthquakes, wars, genocides, and epidemics we know seem much more real. Car accidents and homicides alone kill tens of thousands Americans a year—not to mention diseases like cancer and heart attacks. Do we really have to worry about meteors too? We do worry about lesser dangers, such as being hit by lightning, dying in an airplane crash, burning to death in a fire, being bitten by a poisonous snake, or dying of food poisoning.

Averaged over time, how does the calculated mortality rate from an asteroid or comet impact compare with that from more familiar perils? David Morrison of NASA's Ames Research Center and his co-workers have performed and popularized detailed calculations. According to Morrison, the chance of dying from a Tunguska-class impact is about one in 30 million (per year) and from a global impact catastrophe, one in 2 million. Another way to look at the odds is to calculate the average number of annual fatalities from impacts. Over long periods of time and considering all sizes of event, about three thousand people a year can be expected to die worldwide from asteroid and comet strikes. (In most years the number that die is zero; rarely, very rarely it will be in the millions or even billions.) For the United States alone, the average number is about 150—more people than die from tornadoes and many more than are killed by lightning, snake bites, and food poisoning combined annually. One hundred fifty is also about the number of Americans that die every year in commercial airline crashes, and the same number electrocute themselves at home. Our government spends many millions of dollars tracking violent storms, protecting the food supply, and making airline travel safe. Should we not look further into the impact hazard? The chances of such a disaster today, tomorrow, or even in the next few centuries may be small. But no other threat, except perhaps nuclear war, could destroy the world as we know it.

Protecting the citizens of Earth against cosmic crashes would require a three-pronged effort. First, we must survey the skies and detect as many Earth-approaching comets and asteroids as possible. Second, we must develop the capability to accurately track any object that could conceivably be a threat, so that we know exactly where on Earth and when it will strike. Third, if we hope to prevent the crash rather than merely evacuate the inhabitants of a strike zone, we must develop the means to intercept an oncoming cosmic projectile and deflect it from its course. Although all of this has a science-fiction ring to it, our existing technologies may be adequate to substantially reduce the risk.

Of the projectiles that could menace our planet, about 90 percent are near-Earth asteroids or short-period comets. The rest are longer-period comets that return at intervals greater than twenty years. Planetary astronomers estimate that there are about two thousand Earth-crossing asteroids larger than one kilometer in diameter. Among the more than 130 Earth-crossing asteroids currently catalogued, none have a present orbit that will lead to a collision with Earth in the next few centuries. But a near-encounter with one of the planets, such as Jupiter, could perturb a safe orbit into a deadly one. Large Earth-crossing asteroids are hard to find largely because their poor ability to reflect sunlight makes them very dim. Some of these dark moonlets could be on a dangerous course, although if we find one that is, we will likely have decades to do something about it.

With current techniques, astronomers are finding several Earth-crossers each month. Based at the University of Arizona, the Spacewatch system uses a 0.9-meter wide-field telescope with an electronic scanning camera to find asteroids in real time. With its extensive use of automation and sophisticated computer software, Spacewatch has much in common with robot telescopes, used to find distant supernovae. Unique to Spacewatch, however, is its ability to operate continuously rather than compare time exposures made separately. Spacewatch is now finding about half of all new near-Earth objects, including some as little as 10 meters in diameter. Yet to thoroughly understand the behavior of asteroids and short-period comets that pose hazards to Earth, much better detection capability will be needed. Larger telescope apertures would help find dimmer objects and objects farther away. Spacewatch observers are planning to double the size of their telescope and much increase the span of their electronic detector array. But finding most of the thousands of large Earth-crossers within the next few decades rather than centuries will require a much larger-scale project. NASA is considering a proposal to build six or more large dedicated telescopes to scan the entire sky. Even if a long spell of cloudy weather crippled one telescope's vision, another would always be on guard. This proposed Spaceguard system could detect about five hundred near-Earth objects and a hundred thousand main-belt asteroids a month.

When a short-period object is discovered, we have the leisure to watch it make multiple passes around the Sun while we refine orbital calculations and figure out what to do about a possible impact. By contrast, with a long-period object we do not. Previously unknown long-period comets appear unpredictably in the outer planetary gloom, streaking our way. Since they are as likely to be orbiting the Sun opposite to the Earth's direction as with it, their potential impact speeds are even greater than those of short-period projectiles. Their usually large size—4 kilometers and up—makes them still more hazardous. These Earth-crossing comets only become visible as heat from the Sun begins vaporizing their long-frozen ices, usually in the vicinity of Jupiter's orbit. About a year of acceleration remains before they swing around the Sun or, rarely, collide with a planet. About half of all long-period comets are actually Earth-crossers—that is, they come closer to the Sun than one AU. If we are especially unlucky, a new comet on collision course with Earth could be detected with only two months remaining before the fatal crash. Spaceguard, with its full sky coverage and immunity from bad weather blackouts, would give us a much better chance to discover a dangerous comet early in its fall through the solar system.

Using optical telescopes alone, establishing the orbit of a distant asteroid or comet accurately enough to pinpoint the location of ground zero and time of Earth impact is difficult, perhaps impossible. Fortunately, astronomers have a powerful tool to track such objects once they have been detected—radar. The great radio telescopes at Arecibo, Puerto Rico, and Goldstone, California, make excellent planetary radars and could show us the size, shape, and surface features of an invader, perhaps even its rotation. Its trajectory could be refined to great precision. Computers might then guide an intercepting spacecraft close enough to the object for onboard sensors to lock on, much as a guided missile launched from an airplane or ship locks on to its target today. So highly developed is today's technology for guiding missiles and interplanetary space probes that a mission of this nature seems but a small leap beyond our present capacity.

In one plan NASA is considering, at least two spacecraft missions would be sent to an oncoming cosmic projectile. The first mission would be purely for reconnaissance. The small spacecraft could fly by an Earth-crossing object, rendezvous with it, or even land on it. The second spacecraft would be larger, armed with nuclear explosives intended either to divert the projectile or blow it up. In order to decide on a strategy, mission commanders would need to know what the asteroid or comet is made of. Will it fragment easily? If it is a comet, could a small blast trigger jets of cometary gases? Such jets are sporadic in nature, but comets may change their trajectories significantly as a result of them.

Early detection is crucial. It is much easier to deflect an object from Earth collision when it is far away than when it is close. At a great distance, only a small change in the object's velocity must be induced, which takes a lot less energy. The best place in an asteroid's orbit to give it a kick is when it rounds the Sun at its closest distance of approach (perihelion). A small impulse there produces the biggest change in position when, months or years later, the asteroid reaches the vicinity of Earth. For an asteroid with an exactly known orbit, changing its velocity by only 1 centimeter per second on the other side of the Sun could be enough to transform a potential impact into a near miss.

With current technology, the only means we have to deliver a hard kick to an asteroid or comet is a chemically powered spacecraft carrying a powerful explosive. For a large asteroid, such an interceptor could either deliver a large nuclear bomb to its surface, bury a charge inside, or explode a warhead some distance away. For smaller asteroids, those less than 100 meters in diameter, we might get by with conventional (nonnuclear) explosives detonated at a great distance. All of these methods would work by blowing mass off the surface of the threatening asteroid; the resulting recoil would knock it off course. A surface blast could produce greater deflection than one at a distance, but a near-Earth intercept, which requires a big blast, might fragment the onrushing object into many pieces. Some of these pieces might remain on collision course with Earth and be large enough to produce global catastrophe. Great backup resources would be needed to deal with such an eventuality. Burying the explosive is more efficient than a surface blast but is riskier still. A distant blast provides less deflection but is more predictable, since there's much less chance of the asteroid or comet breaking apart. For a comet it would be more difficult to tailor any burst with predictable effect because the nucleus is hard to see and because outgasing jets might cause suprising changes of orbit.

In October 1993, Harvard astronomer Brian Marsden put out a warning about a periodic comet known as Swift-Tuttle. Discovered in 1737 by a Jesuit missionary, this over-10-kilometer heavyweight made two later passes through the inner solar system—in 1862 and in 1992. Marsden calculates a one-in-ten-thousand chance for Swift-Tuttle to hit Earth when it returns again in August 2126, because jets on the comet's surface could alter its course unpredictably. But analysis of the comet's orbit since 1737 indicates that jets have so far played little role. Donald Yeomans of CalTech and NASA's Jet Propulsion Lab has calculated that the closest distance of Swift-Tuttle's approach will be 14 million miles on August 5, 2126.

Even if we had only a few weeks of advance warning, a powerful enough nuclear blast could probably push an asteroid or comet far enough off course that it would miss the Earth. For a large, fast comet that would collide on its first pass through the solar system, a chain of explosions might be necessary to implant a final huge blast deep within it. With more warning, we'd have the luxury of giving an object a kick, checking its new orbit, giving it another kick if necessary, looking again, and so on. This strategy could reduce the energy necessary for successful deflection and therefore the size of the intercepting spacecraft. For decades, rocket engineers have talked about developing nuclear rockets. With their much lighter weight compared with giant chemical rockets, they would make superior interceptors, but the development costs would be high and the lead time long.

Another approach to asteroid defense (favored by those who oppose any use of nuclear energy) is to deliver a large rocket engine to the surface of a doomsday asteroid and then fire it up. If such an engine were delivered early enough, it might spare us the need to use nuclear weapons.

But since nuclear energy produces a million times more energy per kilogram of fuel, we'd probably turn to it in the face of a real threat. NASA and Los Alamos scientists have been quick to point out the irony that "the only energy sources that could keep us from sharing the fate of the dinosaurs might be the ones that brought us to the brink of extinction in the cold war."

Controversial among interception experts is the need to respond to the threat of small asteroids in the Tunguska class. Since they are more likely to hit us but are easier to deflect, some argue that we should hone our skills on them. A hit by a 100-meter asteroid in a densely populated area would be so devastating that developing deflection technology is worth considering, whatever the cost. Since these objects arrive at a rate of about one (or more) per human lifetime, we wouldn't have to wait centuries to find out if interception and deflection really worked. Assailants up to about 70 meters across heading for a city-busting collision could be deflected into the ocean without any explosives at all—a massive intercepting spacecraft could just slam into it and knock it off course. Unfortunately, the Spaceguard system could not detect small asteroids until they were only weeks (or less) away. That means we'd need to keep our interceptors on constant alert—an expensive proposition.

Other experts, especially Clark Chapman and David Morrison, argue that only a civilization-destroying 1-kilometer-plus asteroid or comet is worth gearing up our defenses for. We'd have many years warning before a large killer asteroid hit, so we wouldn't need to prepare a defense until we were sure that an impact was inevitable. But this argument ignores the threat from long-period killer comets, for which we'd have little warning. Scientists from the Los Alamos and Livermore national labs, where the Defense Department's Star Wars (Strategic Defense Initiative) research has been concentrated, seem to be pushing for early space experiments. By contrast, academic scientists want to emphasize detection strategies and put off defense work until more of the technical issues are resolved. Both sides may be responding primarily to self-interest: Space warriors want to gear up for war in space, even if killer rocks rather than Soviet missiles are the enemy, while astronomers would like to see more money spent on telescopes.

Tracking Earth-crossing objects solely for the purpose of protecting lives seems to us like a dubious investment. There are many other dangers to human life (from poverty to disease to war) whose prevention might be less expensive. The Spaceguard early warning system, with six two-meter telescopes, would cost about $50 million to set up, with $15 million a year in operating costs thereafter. Its supporters claim that it could cut the risk of *unforeseen,* sudden impacts in half within a decade, and by four in two or three decades. Just detecting a cosmic assailant could allow us to minimize its damage greatly, even if we didn't try to defend against it.

On the other hand, the Spaceguard project could also be justified on purely scientific grounds. It would give an enormous boost to our knowledge of comets and asteroids and therefore to the history of our solar system. If an incoming object were discovered zooming in on a collision course with Earth, we would have several options. If we had no capability to deflect it and the object were relatively small, a mass evacuation of the target region could be planned; most likely there would be many years to execute such a plan. For the long-period comet, however, the warning would probably be less than a year. Finally, if the object were large enough to threaten global catastrophe but we had decades of warning, we might still have the opportunity to develop and test interception and deflection technology before the potential collision. If a long-period comet appeared on collision course with Earth, however, we would have no time to develop such a capability. Shouldn't we develop it now?

Putting a lot of effort and money into nuclear weapons and related research now, just as the cold war is winding down, doesn't feel right to many well-informed people. Many billions were wasted on Star Wars research during the 1980s; it is fair to ask whether we really want to go a similar route now. The national labs seem to be making a successful transition into a post–cold war era of peaceful research; a new emphasis on nuclear explosions might abort that transition.

Carl Sagan has expressed his concern that the very technology that could be used to deflect a menacing asteroid from hitting Earth could also be used irresponsibly to force a harmless one onto a collision course. He asks whether we really want to create a technology that could thus cause global catastrophe. "Can we humans be trusted with civilization-threatening technologies?" he writes. The probability of global catastrophe due to impact is somewhat less than one in a thousand per century—which makes it unlikely that the ability to steer asteroids will get into some madman's hands in the next hundred years. Now only the United States and Russia have enough nuclear weapons to unleash megadeath. Asteroid steering could give this ability to many nations—and many madmen—inexpensively.

It seems to the authors that solving the technical problems of pushing asteroids and comets around would require a vast effort of enormous cost. It would be *more* difficult and more expensive even than building a nuclear capability sufficient to devastate a city. Despite the current lull in nuclear posturing, we are still on the brink of global catastrophe and will remain there until nearly all nuclear weapons are destroyed.

CHAPTER TEN

Impacts and Evolution

Once more than sixty species of dinosaurs, large and small, herbivores and carnivores, lived on Earth. Paleontologists have unearthed remains of more than five thousand individuals; from babies in their nests, to complete skeletons of Tyrannosaurus rex, to bones belonging to a 120-foot-long plant-eater called Ultrasaurus. In body form and function the dinosaurs were as diverse as modern mammals—and were every bit as successful at survival. Some had incredibly long necks with tiny heads, or ducklike bills studded with sharp teeth; others had bony plates and spiked tails; still others had long curved claws and a huge head filled with teeth the size of daggers. Whatever their form, dinosaurs lived just about everywhere on Earth, even in what is now Alaska. Their reign lasted for more than 150 million years.

The popular myth that dinosaurs were overgrown clumsy creatures ill adapted for survival may finally have been put to rest by the movie *Jurassic Park.* By analyzing the spacing of dinosaur tracks, specialists have shown that some of the predatory species could run quickly for substantial distances. For this and other reasons, they concluded the creatures that left the tracks were warm-blooded and as active as modern large predators. Extrapolating far beyond current research, the makers of *Jurassic Park* depicted predators as not only active but curious and diabolically clever. One doesn't need to believe that dinosaurs had intellects rivaling ours to appreciate that they did not die in the mass extinction of 65 million years ago because something was wrong with them, because they were unfit or genetically exhausted, as earlier theories suggested. Today's large predators, such as lions, wolves, and bears, might have died out, too, if the food chain had collapsed under them.

One of the most remarkable consequences of the idea of the *bangs* that devastate Earth is the radical change they have made in our understanding of evolution. Central to the theory of evolution proposed in 1858 by Charles Darwin (and independently by A. R. Wallace) is the notion that species change by a gradual process of natural selection, also called "survival of the fittest." Innumerable small variations exist between individuals, some of which can be inherited. Individuals whose variations make them better adapted to their environment—for example, individuals that hunt better, find more grass to eat, or swim away from their enemies faster—survive to have more offspring. They pass their characteristics on to later generations with greater efficiency than their less well adapted competitors. In each population, the proportion of individuals with a particular beneficial trait gradually increases, while that with a harmful trait decreases. Over long periods of time, such natural selection allows a species to change in form and function. A key survival benefit of this mechanism is that species can adapt to changes in their environment, if the changes aren't too great and don't happen too suddenly.

During the nineteenth and early twentieth centuries, biologists, paleontologists, and geologists (including Darwin himself) uncovered an enormous body of fossil and geological evidence to support the theory of evolution. They showed that many species no longer alive once existed and that life had changed profoundly over millions of years. For example, paleontologists were able to trace the development of the horse over 50 million years from a dog-sized creature called Hyracotherium to the modern Equus. Later in the twentieth century, with the development of radioisotope and other forms of dating, our knowledge of ancient life improved enormously.

Although most nineteenth-century scientists accepted the fact of evolution, Darwin was unable to convince them of the role of natural selection. His attempts at persuasion were hobbled by his ignorance of how inheritance really works. By 1865, Gregor Mendel had discovered and published the laws of genetics, which explain how traits are passed from one generation to the next. Unfortunately the world was not ready for Mendel's revelations, and his work was forgotten until 1900. Even

Darwin failed to appreciate the significance of Mendel's pea-breeding experiments, which he could have used to clarify natural selection. By the 1940s, biologists had integrated the sciences of genetics and evolution. With the discovery of DNA in the 1950s and the explosive growth of molecular biology came further revelations of how changes within cells allow evolution to proceed.

There has been much disagreement among reputable scientists about the details of how evolution works, but until recently our overall view of natural selection as a *gradual* process remained the same as in Darwin's time. To understand exactly how big bang–caused mass extinctions have changed the picture, we'll need to delve a little further into the mechanics of natural selection.

Most of the variability in organisms that reproduce sexually arises from the nearly infinite number of gene combinations inherited from the two parents. (A *gene* is basically a chunk of DNA molecule that determines a particular trait.) Natural selection works on variations, eliminating some while promoting others. In the absence of *new* variations, evolution couldn't go very far; a species would be restricted to the combinations of already-existing genes. Once in a while, however, there is a *mutation*, an abnormal change in genetic material that produces an individual slightly different from those that came before. Many mutations are caused by radiation damage to an organism's DNA (due to X-rays, gamma rays, cosmic ray particles, or other natural radioactivity), others by chemical damage. Some involve breaking of chromosomes, containing thousands of genes, perhaps with abnormal joining of the fragments. Still others, discovered by Barbara McClintock in the 1940s (another example of brilliant scientific work that wasn't recognized for decades), involve pieces of DNA called *transposons* or *jumping genes* that can move from one part of the genome to another. (An organism's *genome* is the totality of its genetic information.) If a mutation occurs in the DNA of a sex cell (gamete), it can be passed on to an individual's offspring. Mutations can also cause cancer.

Molecular biologists have recognized still other mechanisms for generating the variety needed for selection, such as *gene duplication*, an error in DNA copying that results in more than one copy of a gene. But gene duplication is not considered a mutation, since one copy continues to function normally while selection works on the other.

Usually any alteration of genetic material is disadvantageous. If the change is serious enough, the organism inheriting it dies or fails to reproduce. Many changes are neutral and will be passed to subsequent generations with no ill effect. The importance of such neutral mutations to evolution, if any, is highly controversial. Sometimes, however, a mutation gives the individual an edge and enhances its probability of survival. In other words, the altered creature is more "fit." One example might be the first leopard that had spots—its mottled coat made it less visible when it crouched on a tree limb. Beneficial or not, mutations are very rare. For a particular gene, there is only about one mutation per hundred thousand sex cells. The rareness of mutations helps limit the rate at which evolution can normally proceed. Obviously, the rate at which a particular creature reproduces also sets a limit on how fast evolution races along. Bacteria and viruses under stress from chemical attack evolve the fastest; the AIDS virus is known to be a virtual wizard at altering its form to defeat physicians' attempts at drug therapy. Likewise the common cold has defied all attempts to cure it, partly because there are so many fast-evolving strains. Fast-breeding insect species can change their color patterns within years when changing conditions place them under intense selective pressure. At the other end of the spectrum, it usually takes a large animal species millions of years to evolve so much that it is considered to have become a new species. Interestingly, Barbara McClintock reported that the rate of transposition skyrockets when cells are threatened; this makes sense as a way to create the maximum amount of variability upon which selection can work on in a crisis.

Generally speaking, however, the molecular machinery that allows organisms to generate variety cannot run fast enough to respond to catastrophic changes in their environment. That is why mass extinctions due to extraterrestrial impacts force us to rethink evolution. An excessive preoccupation with the question of "fitness" may have distracted scientists from examining the accumulating evidence for mass extinctions. As a result, we now think it likely that they have been misled for more than a hundred years. They may have fooled themselves into thinking the primary driving force of evolution has been competition among individuals and species under *ordinary* circumstances, when in fact the driving force has been another phenomenon entirely.

In case after case, the fossil record reveals a group of organisms thriving over a wide geographical range. Then they just vanish, sometimes forever. For example, along with the dinosaurs and the forams, a large group of shelled mollusks called ammonites disappeared. They had lived in the oceans worldwide. Some ammonites resembled today's beautiful chambered nautilus. Some were almost a meter in diameter, but most were far smaller.

Such mass extinctions may be necessary to all major shifts in the direction of evolution. Indeed, as we mentioned in the previous chapter, paleontologists such as David Raup suggest that extraterrestrial impacts are the main cause of mass extinctions. But if they are right, it is *asteroids and comets that supply the main driving force of evolution,* not gradual natural selection. In other words, impacts produce such rapid change in conditions that creatures of many species are suddenly no longer "fit" to survive. In the absence of enough variety or any means of producing it quickly, no individuals of these species can adapt to the new conditions, so those species die out. No animal species can adapt in the evolutionary sense *during* the catastrophe, but as Raup puts it, "some species are fortuitously *pre-adapted* to impact effects" and therefore do survive. In other words, characteristics that have evolved for other reasons happen to protect them. After the dust settles and the Sun emerges from the gloom, the few surviving species proliferate rapidly. On this newly "open field" with minimal competition, these founding species ultimately give rise to many more species. Thus natural selection continues operating through the mass extinction—not gradually but at a vastly accelerated pace.

Some reputable paleontologists still read the fossil data differently and find that mass extinctions occur over millions of years, not suddenly; or they still insist that impacts have only local effects. But paleontologists such as Raup and Stephen Jay Gould gently suggest that such resistant positions are the result of gradualist bias, not objective analysis. In their view, the history of life on Earth is one of long periods during which species change slowly if at all, punctuated by bursts when new species flower. Violent impacts may be the punctuation marks. We are certainly not claiming the existence of evidence that impacts caused other mass extinctions, to match the evidence that a comet or asteroid killed off the ammonites, forams, and dinosaurs. Much research still needs to be done, and it is being done, both on mass extinctions and on crater geophysics. But what a fascinating and revolutionary vista the impact research has opened up: Physics intervenes twice in the development of life, first by the random mutations necessary for all evolutionary change, then by wreaking destruction on a global scale so that creatures that otherwise would have had no chance can prosper.

After dinosaurs departed from the scene, the relatively few and previously unimpressive mammals began a spectacular expansion. Every single dinosaur species was wiped out (if we don't count birds). However, many mammalian species survived. Thus humans, as the latest in the long line of mammals, owe their very existence to the big bang that devastated life-forms on Earth 65 million years ago.

Just as catastrophic astrophysical events played a part in our origins as a mammalian species, they mediated the birth of our very atoms. But the scale of violence, the temperatures needed to cook primeval matter into forms necessary for life, were far greater than those produced by comet crashes. As difficult as it has been to demonstrate the role of extraterrestrial causes in influencing our evolutionary heritage, it has been even more of a struggle to tease out the secrets of our nuclear pedigree. Three tumultuous centuries of progress in physics and chemistry were required just to frame the key questions: What are the elementary particles of nature? How, where, and when did the chemical elements form? It took thousands of years of astronomy even to locate the site where matter evolved into its present state of complexity. And now we think we know. It happened in the heart of a star.

II

Exploding Stars

CHAPTER ELEVEN

New Star

On February 22, 1987, in two giant tanks of water—one in a Japanese lead mine and the other under Lake Erie in a salt mine—nineteen blue flashes of light automatically triggered electronic detectors. Never before had these detectors recorded so many flashes in such a short time. No human would confirm them for days, but they recorded an outburst that had occurred 175,000 light-years away. In a burst lasting less than twenty seconds, vast numbers of neutrino particles from within an exploding star had traversed the underground detectors; a few had collided with particles in the tanks, producing flashes of light called Cerenkov radiation.

About midnight on the evening of February 23, Oscar Duhalde peered up at the Large Magellanic cloud, a neighboring galaxy in orbit around our own Milky Way galaxy. A skilled assistant at the one-meter telescope at Las Campanas Observatory in Chile, Duhalde was very familiar with this part of the sky. Within the Large Magellanic cloud, he noted the faint patch of glowing gas called the Tarantula nebula. But right next to this nebula, Duhalde saw a bright spot he had never seen before.

A few hours later, astronomer Ian Shelton, working with a smaller telescope on the same mountaintop, developed a photographic plate that he had just made of the same section of sky Duhalde had observed. Shelton was startled to see a sizable spot just to the southwest of the Tarantula nebula. His plate from the previous night showed only a very dim star at this location, yet the spot he saw now corresponded to a star bright enough to be seen without a telescope.

Shelton, Duhalde, and several colleagues went outside for another look. The new star was still there. A few moments' reflection convinced the astronomers that, based on the distance of the Large Magellanic cloud from Earth, the new object could only be an exploding star, or supernova, just beginning its rise to maximum brilliance. In two thousand years of recorded observations of the sky, only six supernovae had shined brightly enough to be seen with the unaided eye. And the last one had been in 1604, before even the invention of the telescope! Thus began the most sensational astronomical saga of the 1980s, a stunning example of our second big bang; it would not be topped in the popular imagination until comet Shoemaker-Levy 9 assaulted Jupiter in 1994.

An hour after the discovery in Chile, New Zealand amateur astronomer Albert Jones turned his telescope on some variable stars in the Large Magellanic cloud. He too saw the bright new star that didn't belong where it was. Clouds frustrated his attempts to measure the new star's brightness, so Jones called colleagues in Australia and New Zealand. After the skies cleared, he continued observing and was able to chart the rising brightness of the supernova for several hours. Alerted by phone, Australian astronomer Robert McNaught realized he'd taken a photo that might have revealed the supernova the night before, but hadn't checked it yet. The new star blazed in that photo, too, at a lesser brightness than he could see now, but still clearly there.

Credit for a discovery in science, vital to building a reputation, goes to the first person confident enough of what he has observed to make the discovery public. In observational astronomy, that usually means the first astronomer to call Brian Marsden, who runs the Central Bureau for Astronomical Telegrams of the International Astronomical Union, or IAU, in Cambridge, Massachusetts. At about nine A.M. on February 24, Marsden received a telex from the Las Campanas Observatory reporting the supernova. Minutes later, he received a call from McNaught, reporting the latest brightness measurements. Marsden's sleuthing soon established that Albert Jones had made the only independent discovery of Supernova 1987A, although it occurred a few hours after the discovery in

Chile. Officially, credit for the discovery of Supernova 1987A went to Shelton and Duhalde.

Supernova 1987A set off a wild rush by astronomers worldwide. The brilliant outburst offered a once-in-a-lifetime opportunity to make detailed observations of one of the most enigmatic phenomena in science. Witnesses at one time assumed that the flare-up of a new spot in the sky was a star's birth; today we know such an appearance is more likely to reveal a star's death. Supernovae mark the cataclysmic end of stars, a phenomenon of central importance to astrophysics. In the case of Supernova 1987A, the deceased was an obscure faint star previously known as Sk -69 202. This star had a spectrum characteristic of a blue supergiant, a massive star some fifty times the radius of the Sun. When astronomers went looking for Sk -69 202 after the discovery of Supernova 1987A, they found it had disappeared.

Supernovae reveal crucial information about the life cycle of stars, but their importance to our story of origins lies more in this key fact: They are the only source of many chemical elements necessary for life. Thus supernovae explosions qualify as our second big bang. As we shall see, supernovae may also influence the evolution of the universe by providing an energy source to trigger the formation of stars. And they almost certainly accelerate the high-energy cosmic rays that cause many of the mutations needed for the evolution of life. Because some types of supernovae evolve with standard brightness, they may also help establish the age and fate of the universe.

Everything about supernovae is fabulous. Many of them mark the explosion of stars far more massive than our sun. The violence required to rend a massive star apart is almost beyond comprehension. During the first seconds of its explosion, a single supernova radiates energy at a rate comparable to that of the entire universe, which contains at least 10^{21} stars shining by thermonuclear fire. Supernova explosions create some of the most bizarre objects ever conceived—whirling neutron stars made of matter so dense that a teaspoonful would have a mass greater than ten battleships. The most mammoth explosions can produce black holes, or invisible stellar remants whose gravity is strong enough to prevent any light from leaving and to trap forever any matter that comes too close.

None of this was known or even suspected when the first historical supernovae were observed centuries ago. It is conceivable that the bright but temporary Star of Bethlehem reported in the Bible was a supernova.[1]

Chinese and Roman records from the year A.D. 185 declare that a new star in the constellation Centaurus shone for twenty months. At its peak, the star was easily visible by day. Chinese records from the year A.D.393 report a similar new star. Modern astronomers have linked these events to patches in today's sky that we know as supernova remnants RCW 86 and CTB 37 A/B.

Even more spectacular was the supernova of 1006. First noted by Japanese, Chinese, and Egyptian astronomers, this most brilliant of recorded supernovae brightened until it outshone Venus and all other planets and even rivaled the moon. It was visible by day for months, by night for almost three years. Ultimately, witnesses chronicled its appearance all across Europe and North Africa. Left behind was an expanding shell of gas detectable today as the radio source PKS 1459-41, as a weak X-ray source, and as some faint wisps visible through a large telescope.

Oddly, the next supernova, the famed event of 1054, was apparently not recorded anywhere in Europe. Chronicled with care by the Chinese and probably observed by peoples of the American Southwest, this "guest star" in the constellation of Taurus blazed by day for three weeks and at night for nearly two years. It bequeathed the beautiful expanding remnants we know as the Crab nebula. At its center, a spinning neutron star beams radio pulses into the galaxy at a rate of 30 pulses per second. This *pulsar,* discovered in 1968, is slowing down slowly due to energy loss, at a rate consistent with its age of about a thousand years. Rare among pulsars, the remnant neutron star in the "crab" also gives off pulses of visible light.

Seen in the night sky for six months, the supernova of 1181 was recorded only in China and Japan. Its remnant is the strong radio source

1. More likely, if the biblical account is correct, it was a more common and shorter-lived nova. Novae are relatively minor flare-ups believed to occur when hydrogen escapes from a star and falls onto its white dwarf companion. Hydrogen builds up until it suffers an explosion similar to a thermonuclear bomb. Such stars survive and may even blaze with repeating novae at regular intervals. Unlike supernovae, visible for the better part of a year or more, novae typically shine brightly for only days or weeks.

3C 58. There's no pulsar associated with 3C 58 or with PKS 1459-41, because the beams given off by the remnant neutron stars happen to miss us or, possibly, because neutron stars do not remain.

In 1572, the next bright supernova—or in Latin, Nova Stelli—appeared just in time to play a crucial role in the history of ideas. For thirty years thinkers had heatedly debated Copernicus's bold and dangerously heretical theory that the Earth is but one of several planets circling the Sun. The opposing view, originally from Aristotle and sanctified by the Catholic Church, was that a motionless Earth occupied the exact center of the Universe. The Sun, Moon, and planets spun around it at differing rates in various "crystalline spheres." Stars occupied an eighth rotating sphere of the heavens. In contrast to the changeability and imperfection of the lower spheres, the eighth and highest was invariable and holy. There was no place for a temporary star in the eighth sphere. To the defenders of Aristotle's astronomy, all such flare-ups, along with comets and meteors, were atmospheric and therefore unimportant. This notorious bias may account for the failure of Europeans to record the supernovae of 1054 and 1181, which should have been clearly visible to them (although they did report the supernova of 1006).

The young Danish astronomer Tycho Brahe was not the first to notice the brilliant new star in the constellation of Cassiopeia. But he made detailed observations that enabled modern astronomers to reconstruct its light curve, or graph of brightness versus time. Even more important, he measured the new star's position with respect to background stars. Brahe found that there was no change in this position whatsoever from one night to the next. By contrast, the Moon, planets, and comets show an easily detectable motion with respect to the stars from night to night. Brahe's discovery therefore located the supernova firmly within the eighth sphere, where the Aristotelians couldn't explain it.

In a controversial book entitled *De Nova Stella,* Brahe used his measurements to argue against the Aristotelian crystalline spheres. While this one observation was not enough to bring down the Aristotelian system, it did raise profound doubts in his more open-minded contemporaries. What was more, the new star of 1572 inspired Brahe to his life work of careful observation of the planets. His data led Johannes Kepler to discover his famous laws of planetary motion. When Isaac Newton explained Kepler's laws with his own laws of motion and universal gravitation, the Aristotelian position was decisively undermined. By the mid-eighteenth century, Copernicus and modern science were victorious.

In 1604, Europeans were startled by another supernova, the last one visible to the unaided eye until 1987. This time the new star appeared during and very close to a near-conjunction (appearance at the same spot in the sky) of Mars and Jupiter, which was thought to have great astrological significance. Tycho Brahe's protégé Kepler published a book on the 1604 nova in which he again pointed out the apparently great distance of the event (based on the fact that it didn't move with respect to the stars), in contradiction with Aristotle's notion of changelessness in the higher regions of the sky. Within five years Galileo, who also observed the event of 1604, had constructed his telescope and begun making observations that further weakened the Aristotelian worldview. Considering the events of 1572 and 1604, it seems fair to conclude that supernovae have profoundly influenced our intellectual history, not to speak of our physical origins.

Each of these historical "new stars" shone long enough that they must have been supernovae rather than ordinary novae. So bright were they that we believe that they occurred within our own galaxy (based on current knowledge of the scale of the universe). Each of them has now been linked with supernova remnants visible with radio telescopes, optical telescopes, or both. Modern astronomers have discovered more than seven hundred supernovae in distant galaxies. In the larger galaxies, they appear to occur at a rate of about one every thirty years per galaxy. We see them at a much lower rate in our own galaxy only because interstellar dust obscures our view. Beaming over thousands of dusty light-years, starlight dims so that most supernovae—and for that matter, most ordinary stars in our galaxy—cannot be seen from Earth.

Neither observers of historical supernovae nor the more sophisticated astronomers of the eighteenth and nineteenth centuries had much idea what caused these mysterious outbursts. They certainly didn't realize they were witnessing the deaths of stars. A first breakthrough came via measurements of the velocities of expanding gas in stellar explosions, but real understanding awaited the revolutionary developments of twentieth-century physics.

CHAPTER TWELVE

Stars Are Us

Touch part of your body, or any object lying nearby. According to the discoveries of astrophysics, the matter you touch is a little bit of stardust. That is, it was literally once part of a star. Not just any star, but a star that exploded as a supernova. This intriguing fact has been known since the late 1950s, when William Fowler of the California Institute of Technology and his co-workers presented theories on the evolution of the chemical elements for which he was later awarded the Nobel prize in physics. As the supernova saga unfolded in fascinating detail during the past three decades, physicists became increasingly certain that they had located the main site where matter was "cooked" into its present nuclear form.

We are still not sure exactly how supernovae blow up, but explode they do, flinging nuclear matter from their interior into space. This dust and gas are then incorporated into new "protostars" that collapse and ignite as part of an ongoing process of cosmic recycling. One of these protostars, perhaps after several cycles of star formation and destruction, became our Sun. As we shall see, there is strong evidence that the matter of our solar system, and of life, has been cooked and recooked in supernovae.

About 99 percent of our bodies consists of just six chemical elements: hydrogen, carbon, nitrogen, oxygen, phosphorus, and sulfur. Of these elements, hydrogen and oxygen are by far the most abundant, accounting for more than 85 percent of all atoms in living matter. Although less common at about 10 percent, carbon earns top status in the realm of life by virtue of its ability to bond prolifically with itself and other atoms. Essential also to life are many trace elements, from magnesium, without which plants could not make food; to sodium, essential to our nerves and muscles; to iodine, found in thyroid hormone; and iron, found in blood. Less well known elements such as molybdenum, selenium, and vanadium also play vital roles in biochemical processes.

How did the elements of life, as well as the other ninety or so naturally occurring chemical elements, come to be? Until the early twentieth century, the origin of the elements was a mystery barely accessible to science because the structure of matter itself was unknown. We are not even sure that earlier scientists even thought to ask the question. By the end of the 1920s, however, physicists understood the role of atomic electrons. Comprising a tenuous cloud, these negatively charged particles with little mass orbit a tiny but dense nucleus. After Sir James Chadwick's discovery of the neutron in 1932, it became clear that atomic nuclei consisted of two building blocks, the proton and the neutron, together called *nucleons.* These nucleons combine in more than 2,600 different forms, identified by the number of protons, or *atomic number,* and the neutron number. Species with the same atomic number but varying neutron numbers are called *isotopes.* Only about 280 of these isotopes are stable; the rest, the radioactive ones, decay naturally into other isotopes, with lifetimes varying from tiny fractions of a second to billions of years.

Propelled by the development of the cyclotron (a device that accelerates charged particles) by Ernest Lawrence and his collegues at Berkeley in the 1930s and 1940s, physicists were able to study dozens of nuclear reactions and discover many new isotopes. During and after World War II, the development of nuclear reactors (used to control reactions similar to those that power atomic bombs) made it much easier to study other nuclear processes, especially those begun by neutrons. Scientists eventually were able to simulate conditions in the early universe or in the core of stars, where temperatures in the millions or even billions of degrees prevail. The knowledge we have today of nuclear reactions provides an excellent starting point for theories of how matter evolved in the universe.

Chemical reactions can be spectacular, with violent releases of heat, fascinating color changes, the startling appearance of gooey slime, or the bubbling liberation of explosive gases. But such transformations involve only the atomic electrons; they leave the protons and neutrons of the nucleus quite unchanged. By contrast, nuclear reactions fulfil the alchemists' dreams of transmutation of the elements. They can cause alterations in the number of each of these subnuclear particles, as colliding nuclei merge, target nuclei capture neutrons, or radioactive nuclei spontaneously decay. It is even possible to make gold by successive neutron capture on lighter elements. In extreme cases, such as the annihilation of antinucleons by the nuclei of ordinary matter, nuclear particles can disappear entirely, transferring their energy to the high-energy photons of electromagnetic energy called gamma rays. The energy involved in these changes is typically a hundred thousand to a million times greater than that involved in chemical reactions such as ordinary burning.

Key to element-making are *fusion* reactions, in which two nuclei join to form a heavier one. These reactions liberate abundant energy because the mass of the original nuclei substantially exceeds that of the product. The total number of nuclear particles (nucleons) doesn't change in these reactions. What does change is the *binding energy* of the nuclei; that of the product must be greater than that of the reacting nuclei if a fusion reaction is to give off energy. A nucleus with greater binding energy is more stable and will be formed preferentially in a system of competing reactions.

The immense energy of fusion reactions powers the stars. These reactions are known as *hydrogen burning, helium burning,* and so on. But just as it takes a match to start a fire, some energy must be injected to trigger fusion reactions of colliding charged particles. Since all protons carry a positive electrical charge, they repel one another strongly. Unless the nuclear fuel in a star is very hot, the nuclei will be traveling too slowly to overcome their mutual repulsion. Fusion requires the participant nuclei to get very close together. The reason for this is that the attractive force involved in nuclear reactions, the so-called *strong nuclear force,* has a very short range. Therefore, only at temperatures in the tens and hundreds of millions of degrees are nuclei moving fast enough to penetrate the repulsive barrier so that nuclear fusion reactions can take place. Even then, nuclei get past the barrier only by a quantum mechanical process of "tunneling" through it.

In a fusion or thermonuclear bomb, the necessary high temperatures are achieved by detonating an atomic (fission) bomb. In a young star, the gravitational energy of the collapsing stellar gas cloud provides an initial high temperature. To operate a prototype fusion reactor, which has the potential to generate cheap electrical power, engineers must inject energy to trigger the reactions from outside, perhaps from a giant laser. Today's most advanced reactors produce temperatures that equal or exceed those in the interiors of stars but at too great an energy cost and too low matter density. The trick to making such reactors practical is to extract more energy than you put in. So far, even breakeven operation has eluded fusion engineers.

Physicists reached a basic understanding of fusion reactions during World War II. They realized that fusion provided a natural explanation for the origin of the chemical elements. Most nuclei, at least up to iron, do have greater binding energy per nucleon than the lighter ones. So it is easy to imagine a chain of reactions, each producing heavier and heavier nuclei. The chain can consist of reactions in which a nucleus captures one neutron after another, or it can involve successive charged-particle collisions. But where, the researchers wondered, was the fusion cooker in which these reactions could seethe and simmer? Ordinary stable stars like the Sun are one possibility, but how could the product elements escape from it?

At first physicists thought the fusion cooker was the furiously hot conditions of the early universe. Edward Teller, known as the father of the H-bomb, did some brilliant pioneering calculations. George Gamow and his collaborators showed how hydrogen cooks into helium. They could see how to make tiny bits of lithium, beryllium, and boron, with atomic numbers three, four, and five respectively. But early efforts to devise a complete theory of nucleosynthesis ran into a formidable barrier. There simply are no stable nuclei with atomic weights five and eight. A chain of neutron-capture reactions would be able to make the isotopes of hydrogen and helium but nothing heavier. The chain of charged nuclei reactions staring with proton-proton fusion ends with a bottleneck at atomic mass five. Two helium nuclei could fuse to form the beryllium nucleus ^{8}Be,[1] but such a nucleus decays back to helium nuclei in

1. The superscript 8 in front of the element symbol Be signifies the atomic weight of the nucleus, in this case containing four protons and four neutrons.

only 10^{-16} seconds. How then were elements heavier than helium created? Had nature been unable to solve this puzzle, life would not exist.

Edwin Salpeter of Cornell University and Fred Hoyle (then at CalTech) found the answer. They hypothesized that in the very dense interior of a star, enough ^{8}Be nuclei would react with helium nuclei before decay to form stable nuclei of ^{12}C (carbon-12). Hoyle pointed out that this reaction would proceed fast enough only if carbon-12 could exist in a previously undiscovered excited nuclear state, which decays to form stable ^{12}C. Experiments soon confirmed the potency of this mechanism; the discovery of the excited state Hoyle had predicted was the first of a long series of successes for nuclear astrophysics.

Once the chain of fusion reactions in a star reaches carbon-12, there are no fundamental barriers to producing all the elements up to iron. But the helium-burning mechanism discovered by Salpeter and Hoyle could not have worked in the early universe. One basic problem is that in order to produce heavier and heavier elements, higher and higher temperatures are required. This is true because electrical repulsion between heavier nuclei is greater, since they have more protons; thus higher and higher collision speeds are needed to overcome the repulsive barrier. But in an evolving, expanding universe, the temperature goes *down*, not up. There's just no way to set up a sequence of reactions leading to heavier and heavier nuclei under these conditions. Also, the rate of synthesis of elements such as carbon or oxygen depends on the density of helium present. At the stage of the Big Bang expansion when the temperature allowed such nucleosynthesis (it could not be so hot that nuclei produced would be immediately broken apart), there was far less helium around than in the helium-rich core of a star.

A massive star has a complex structure of gaseous shells, not unlike a giant many-layered onion. At stellar temperatures, the elements in these outer layers are all in plasma form (hot electrically charged gases). Astrophysicists have constructed intricate computer models for nuclear burning in these many-layered stars. Using rates for nuclear fusion reactions measured precisely in the laboratory, they have calculated abundances of the elements. In other words they can predict how much carbon there should be compared with oxygen, how much sulfur compared with iron, and so on. These predictions agree reasonably well with measured solar abundances—exact agreement is hardly to be expected for such hellishly complicated calculations. The original mass of the star greatly influences the mix of elements produced. We don't have much information about the proportion of exploding stars in each range. We also don't know accurately how much of the iron group elements are expelled when stars die and how much become trapped in neutron star or black hole remnants. The agreement is good enough to convince most astrophysicists that the Sun's matter was in fact cooked in larger stars that went supernova.

Another fascinating result of these theoretical studies is the speculation that relatively few supernova explosions of extremely massive stars in the range above a hundred times the mass of our sun might account for most of the observed elements beyond helium. We may even owe our very existence to a single star that exploded 5 billion years ago in our Milky Way galaxy. Had this particular cataclysm not occurred, the matter of our solar system might have contained too little of the heavy elements for life to have evolved—or even for minerals to exist.

One irony of these studies is that heavy elements laboriously built up in the late stages of stellar evolution may be broken up into helium again during the explosion. Fortunately, the calculations show they are formed again in the wake of the expanding supernova shock wave. Another check that we have the right site for nucleosynthesis is that our star-supernova theory requires that the light elements beryllium, boron, and lithium-6 (^6Li) should be very rare since they are bypassed in the chain of fusion reactions. In fact these valuable elements *are* thousands of times less common than carbon or nitrogen—some people have never even heard of them. Their abundances are well explained by a small rate of production in the Big Bang and by collisions of fast-moving protons and helium nuclei in the galaxy with interstellar matter.

What about elements as heavy as iron or heavier? Some of them are essential to life also and fairly common on Earth. At the incredibly high temperatures in the core of a giant star, metals such as chromium, manganese, cobalt, and nickel as well as iron are formed. But high-energy gamma rays break these nuclei apart into helium nuclei almost as fast as they can be made. Since the iron-56 nucleus has the highest binding energy per nucleon of any nucleus and is therefore the most stable, heavier elements cannot be made by the energy-releasing fusion reactions that powered the star until this point. So why do the many elements heavier than iron exist?

Cooking heavy-element starstuff requires something called *neutron capture.* Since neutrons have no electrical charge, they can enter an atomic nucleus without being repelled. Hundreds of different heavy nuclei both stable and unstable can be formed by successive neutron capture, all the way up to lead at atomic number 82. If there aren't too many neutrons around, nearly all the unstable (radioactive) nuclei formed in this way will decay, within a period varying from seconds to months, by giving off a negatively charged electron before they can capture another neutron. Each such decay gives birth to a new nucleus with atomic number one higher than its parent. The stable nucleus thus formed may then capture another neutron, continuing the chain of element formation.

In this way, a long sequence of elements can be built up in evolving stars. Neutrons aren't abundant in stars, because the main fusion reactions don't produce many. So the amount of heavy-element matter formed by neutron capture in stars is much less than the mass of lighter elements. Since considerable time is required, at least compared with the short time scale of an exploding supernova, the neutron capture sequence in stars is called the s-process, where *s* means *slow.* The s-process cannot, however, be the sole explanation of the heavier-than-iron group elements. First, many stable isotopes cannot be made this way at all because the chain of neutron capture and electron decay bypasses them. Second, the abundances of the isotopes that *are* produced by the s-process disagree strongly with measured solar abundances.

The catastrophically brief time scale of a supernova explosion provides a neat way out of this dilemma. This time scale is much shorter than the decay times of heavy nuclei. Thus, in the intense nuclear cauldron of a supernova explosion, a chain of elements can be built up by rapid successive neutron capture. While some of these "r-process" nuclei (*r* for *rapid*) are identical to certain s-process nuclei, many others are different. Between the two mechanisms, nearly all the known stable isotopes can be accounted for—right up to element 92, uranium. The handful of nuclei that cannot be formed by neutron bombardment and beta decay are explained by proton bombardment during the supernova explosion.

Although many details remain to be filled in, scientists think they have unraveled the essential scheme by which matter evolves. Light elements form in the early universe and in stars. Titanic supernovae explosions sprinkled throughout galaxies cook the heavy elements and fling them into space. From the debris of these shattered stars, new stars form.

CHAPTER THIRTEEN

Secret Lives and Violent Deaths of Stars

Supernovae mark a particularly violent crisis at the end of a star's life history, inevitable for some but not for most. Most stars live in a stable but tense balance. Gravity pulls them inward with huge forces arising from their immense mass. The pressure of hot gas, originating from thermonuclear reactions in the stellar core, provides an exactly matching outward thrust. All gases, including those of Earth's atmosphere, exert pressure, due to random collisions of fast-moving atoms or molecules. In a star, temperatures are far higher than those within our atmosphere. Such great heat produces extremely rapid collision speeds, causing thermal pressures capable of withstanding even the crushing gravity of a massive star. Even in the dead, burnt-out hulk of a white dwarf star, there is a balance. Here gravity is opposed by the "force" arising from the exclusion principle of quantum mechanics, the requirement that no two electrons occupy the same state.

Stellar life begins with the gravitational collapse of giant clouds rich in molecular hydrogen. Thousands of these clouds, which also contain dust and helium, are scattered through our galaxy. So vast are these clouds that their masses range from a hundred thousand to more than a million times that of our sun. A typical diameter for these interstellar nurseries is about 100 light-years. Such clouds are cold, dark, and unstable. At their low temperatures, about 10 degrees above absolute zero, there is barely enough pressure to provide support against gravity. Even a relatively small disturbance can trigger an irreversible inward fall. Collision with another cloud can initiate collapse, as can the explosion of a nearby supernova or a density wave traveling through the galaxy. Each of these events produces supersonic shock waves that compress the gas locally, creating clumps in which the balance of forces is tipped toward collapse.

As clouds of gas and dust fall inward, they become *protostars.* When protostars form, their higher density attracts still more gas and dust. Each stage of collapse produces heating, due to conversion of gravitational energy into thermal energy. But heating also increases pressure, which slows the collapse. No one has observed the entire process of stellar birth, but computer models show that it can last from thousands to many millions of years, depending on the mass of gas involved. Since the more massive protostars generate larger gravitational accelerations, they evolve more rapidly. When its temperature reaches several thousand degrees, a protostar begins to glow. Eventually temperatures soar into the millions of degrees, igniting thermonuclear reactions. A star is born.

Once its nuclear furnace has been lit, vastly increased thermal pressure now halts a star's collapse and enforces a state of equilibrium. Thus stabilized, a star may burn quietly for billions of years. Astronomers observe numerous clusters of infant stars, still surrounded by vast hydrogen clouds. The more massive stars burn with greater luminosity (intrinsic brightness) and higher surface temperature. Ultraviolet radiation emitted by these massive stars ionizes the surrounding hydrogen, forming a reddish nebula within a much larger, dark molecular cloud. Some of the most beautiful regions in the sky were formed this way, notably the intensely glowing Orion, Eagle, Swan, Keyhole, Madonna, and Rosette nebulae.

Stars much less massive than our sun will continue burning stably for tens and hundreds of billions of years, far longer than the age of the universe to date. If the universe is closed and doomed to collapse on itself, some of these stars may burn until they are overtaken and perhaps shredded in a "Big Crunch." Regardless of the fate of the universe, the options for which we'll discuss later, the comfortable equilibrium of hydrogen burning in stars is destined to come to an end. Sooner or later, the supply of hydrogen in the star's core will be exhausted. In furiously burning giant stars more than twenty-five times the solar mass, the time of reckoning can come as little as a few million years after stellar birth. By reassuring contrast, our Sun's lifetime will be in the range of 10 billion years, give or take a few billion, depending on which computer model you believe.

For some 5 billion years, hydrogen burning has powered the Sun. But as the end nears, hydrogen in the Sun's core is depleted, ultimately making it more difficult for the stars' outer layers to be supported against the crush of gravity. As these layers squeeze the layers beneath them, compression and the energy from gravitational contraction will cause an increase in temperature. Hydrogen in a shell just outside the former core will heat to the point where it too will ignite in fusion reactions. This heat combined with that from the core contraction will heat the surrounding layers of gas, which will then expand enormously to form a red giant star. In the case of our Sun, its volume will increase to envelop at least the entire orbit of Venus, threatening any life that might then exist in the inner solar system.

When a stellar core heats to 100 million degrees, helium burning begins, forming carbon nuclei. In the case of a relatively low-mass star like the Sun, helium burning will begin about a billion years into the red giant phase. There may be instability and minor explosions in our Sun's distant future, but it will not go supernova. Gradually, as its helium is used up, it will shrink to form a type of burnt-out star called a white dwarf.

For larger stars, the story is far different. What follows is a typical scenario leading to supernova explosion. As each stage of nuclear burning ends due to exhaustion of fuel, contraction causes the ignition of yet another phase of burning, requiring ever higher temperatures to overcome the repulsion of heavier and more highly charged nuclei. Carbon burns to form neon, which itself burns to form oxygen. Carbon and oxygen can fuse to form silicon; oxygen can combine with more oxygen to make sulfur, and so on. Finally, in silicon burning, the ^{56}Fe nucleus of iron is formed. This nucleus is so strongly bound that any reaction with it absorbs energy rather than releases it. For a massive star, as iron forms in its core, the end is near. The doomed star's internal structure resembles an onion, with shells of sulfur and silicon surrounding the core, layers of oxygen, carbon, and helium outside, and an outermost shell of hydrogen.

Amazingly, the final stage of silicon burning in a massive star, which has lived for many millions of years, takes but a day. As iron is added to the core, no further nuclear reactions take place there. The core mass increases, raising the gravitational force to awesome levels, but there is no new heating to counteract it with outward pressure. Only by electron pressure does the core resist collapse. As we mentioned earlier, this form of pressure is required by the quantum theory of physics, according to which no two electrons can occupy exactly the same state. At such a point in the star's evolution, all the space between atoms has been squeezed out. Any further addition of iron to the core will now increase pressure beyond what the electrons can support. Outside the iron core, silicon burning continues, producing one final increment of iron. In a dizzying collapse, the entire iron core of about 1.5 solar masses is squeezed to nuclear densities. Computer models indicate the time required for this process is *less than one second!*

During the collapse, virtually all the electrons disappear, combining with protons to form neutrons. The central part of the core may become a single gigantic nucleus, or neutron star, a few kilometers in radius with an incredible density of about 3×10^{14} grams per cubic centimeter. A spoonful of such matter would weigh as much as roughly ten thousand large ships. What is worse for the star, such material is almost completely incompressible. The rest of the core is still falling inward at a high speed. It bounces on the neutron core and explodes outward in a powerful shock wave. According to computer simulations, the velocity of this shock wave is about 50,000 kilometers per second, or one-sixth the velocity of light. One might ask, if a robot observer of these events (protected perhaps by a stellar diving bell made with materials far tougher than those used to make the submersibles ocean explorers use) could somehow survive the intense gravity and blazing heat of a stellar core, what exactly would that witness see? Since the density of matter in the core is so high, it is essentially impenetrable to light; the observer therefore would see nothing. Since the star's outer layers are unaffected at first, an observer outside the star would for some hours after the core collapse also see nothing amiss.

What happens next is controversial. The shock wave races outward into the various layers of the giant star. It could blast its way outward through all the layers, blowing the star apart and sending most of its mass flying outward into space at enormous speeds. Or the shock wave may stall, while the star's mass falls into the core. In this case, one possibility is the formation of a black hole, as the gravity of infalling matter exceeds the limit beyond which no light can escape. If the star's original mass is large enough, it is also conceivable that a black hole could be formed earlier, in the iron core collapse. Once the shock wave blasts through the surface of the star, it expands enormously, becoming a brilliant sphere with rapidly growing luminosity. As the shock wave penetrates the star's outer layers, heating triggers new nuclear reactions, forming elements heavier than iron and creating radioactive decay products that prolong the explosion. Computer calculations of what happens during a supernova require sophisticated modeling, usually on powerful supercomputers. As theorists' knowledge and the computing power available to them expand, they are able to include more and more details in their models, such as the effects of convection during the explosion. (You can visualize this as huge swirls of irregular heating, roughly comparable to the way swirls of hot air from a heater spread through a room.) Recent calculations show that convection helps the supernova shock wave blast its way through a massive star whose inner layers have collapsed.

Although the shock wave clearly carries a vast amount of energy, the bulk of energy released by the supernova, some 99 percent, is in quite another form. When electrons in the iron core combine with protons, each such reaction liberates an energetic neutrino. *Neutrinos* are particles with a very small or zero mass (physicists aren't sure) that play a crucial role in certain decays. These particles interact very weakly with matter; as a result, they can easily penetrate great thicknesses of matter such as the entire Earth. When a star collapses, a storm of neutrinos flies outward through its layers at the speed of light (or a little slower if the neutrino turns out to have a small mass). As neutrinos burst forth from the collapsing core, the loss of energy causes pressure to plummet still further, hastening the collapse.

Once astronomers had announced Supernova 1987A, scientists at the great underground particle-detection laboratories combed through their data to find evidence for the neutrino storm. They found that hours before the first visual evidence of the supernova, the two most sensitive neutrino detectors in the world both recorded strong bursts of neutrinos.

These detectors, located respectively in a salt mine under Lake Erie and in a lead mine in Japan, consist of huge tanks of water enveloped by photomultiplier tubes. The tubes detect bluish Cerenkov radiation given off by charged particles moving at speeds faster than that of light in water (but still less than that of light in vacuum). Such particles are produced by the one neutrino in about 10^{17} that interacts in the tank. Although only nineteen neutrino events were recorded, this number is what would be expected for a typical supernova at the distance of the Large Magellenic cloud releasing a total energy of about 10^{46} joules. With this amazing observation, modern astronomy has penetrated more deeply into the heart of an exploding star than might ever have been imagined possible. And we've confirmed the idea that a star's core can collapse.

It is ironic that a collapse can lead to an explosion. Something similar can happen when a TV picture tube is smashed. Air pressure outside the tube is much greater than inside. Fragments of glass are pushed inward at first, but some may ricochet dangerously. Here it is obvious that the source of energy for the "explosion" is moving air molecules. In the case of a collapsing star, the energy source is gravity—the gravitational energy of the star's outer layers. Supernovae believed to form by collapse are known as Type II. Since the parent stars have outer layers of unburnt hydrogen, astronomers expect to see spectral lines of hydrogen when they look at Type II supernovae. Astronomers usually see such supernovae in the arms of spiral galaxies, known to be rich in younger massive stars. But many supernovae show no hydrogen lines. If they weren't formed by collapse, what caused them to blow up? Astronomers believe their parent stars are white dwarfs.

Unlike massive stars, white dwarf stars are very common. As mentioned earlier, they are the burnt-out remains of stars about the mass of our Sun. They lack hydrogen; it has all been consumed. Nuclear reactions no longer provide energy within them, but some still have enough heat left over from their glory days to glow faintly. Sirius B, one of the nearest stars to our solar system, is a typical white dwarf. The matter inside white dwarfs is said to be "degenerate"—that is, its huge pressure comes not from heat but from electrons in a collapsed state very different from that in ordinary atoms. White dwarf matter is so dense that a spoonful would weigh many tons on Earth. Left to itself, a white dwarf would cool over billions of years until it stopped glowing and its temperature approached absolute zero.

White dwarfs would be relatively boring and not much use to supernova theorists were it not for the fact that many are part of *binary* (two-star) systems. In some cases the two stars revolving around each other are too far apart to exchange material. In other binary systems, however, a significant amount of mass can fall from its stellar companion onto the white dwarf. This phenomenon, known as *mass accretion,* is especially likely if the companion becomes a red giant. Astronomers detect some binaries when one star eclipses the other. The periods of revolution, which can be as little as a few hours, indicate that the stars are close enough to twist each other's shapes by tidal forces. With such small separations, it's easy to understand how one star could rip material from the surface of its companion.

Mass falling onto a white dwarf creates the opportunity of new life for the star, but it also sets the stage for a possible violent death. Hydrogen or helium can form a surface layer in which thermonuclear reactions ignite. This burning can proceed explosively, leading to ejection of a shell of hydrogen; such is the cause of common novae formerly confused with supernovae. Novae do not disrupt the interior of the white dwarf and can reoccur many times. However, there is a limit to the amount of mass a white dwarf can accept. The mass limit of about 1.4 solar masses was discovered by University of Chicago astrophysicist Subrahmanyan Chandrasekhar. Above the *Chandrasekhar limit,* the pressure of degenerate matter can no longer support the star's mass.

Once a white dwarf picks up enough mass from its close binary companion to exceed the limit, it is doomed. Pressure in the interior rises, leading to a skyrocketing temperature. Carbon and oxygen nuclei fuse at a runaway pace. Because most of the matter is still degenerate, the star cannot expand gradually and burn stably. Instead, fusion reactions proceed rapidly through the silicon-burning stage in a tremendous thermonuclear explosion. The result is a Type I supernova. There is little or no trace of hydrogen. Despite some astronomers' optimism that Type I supernovae have been explained, there are major mysteries: What is the exact nature of the explosion? What is nature of the mass transfer or merger? Why has no white-dwarf-containing binary star system been discovered that is both small enough to merge in the age of the universe and massive enough to trigger a supernova explosion?

Both types of supernovae, the Type I that lack evidence of hydrogen and the Type II that have it, cause spectacular flare-ups in the sky that humans have marveled at for thousands of years. Neither of these mind-boggling big bangs have yet been totally explained, but astronomers are confident they are on the right track. Type I explosions, which demolish white dwarf stars of a constant 1.4 solar masses, seemed (at least until very recently) the better bet for taking the measure of the universe and determining its fate. And the Type II outburst shatters far larger parent stars.

CHAPTER FOURTEEN

The Peculiar Progeny of a Supernova

Of all the beautiful sights in the night sky, there is one object that has taught us the most astrophysics—the Crab nebula. At a distance of 6,300 light-years, it lies well within the Milky Way galaxy, in the next spiral arm outward from our own. The Crab is the clearest and best-studied supernova remnant, the fragments of a massive star that died in the spectacular outburst witnessed in the year 1054. Not so long ago, its tattered filaments, diffuse central gases, and strong radio and X-ray emissions baffled astronomers. Nicolas Mayall once borrowed Winston Churchill's famous description of the Soviet Union to call the Crab "a riddle, wrapped in a mystery, inside an enigma." Much as historical events have raised the Iron Curtain, a series of brilliant discoveries has transformed the Crab nebula into our best-known example of supernova dynamics.

In 1745 a wealthy English physician and amateur astronomer named John Bevis spotted a faint patch of light in the constellation Taurus. Too dim to be seen with the naked eye, this diffuse cloud is about one twenty-fifth the apparant size of the Moon. Charles Messier, who published the first catalog of nebulous objects, discovered it independently in 1758. In modern photographs taken with high-resolution telescopes, the nebula doesn't look much like a crab, but to Irish observer William Parsons, the third Earl of Rosse, a prominent gap in the nebula suggested the cleft between two claws. Lord Rosse not only named the nebula, in 1844 he was the first to draw its wispy filaments. By the 1920s, measurements made years apart revealed that the filaments were expanding outward at great speeds. By the 1940s astronomers had linked the Crab Nebula with the Chinese "Guest Star" of 1054 and suggested that a supernova explosion was responsible for the filaments. When the measured expansion of 1,500 kilometers per second was traced backward (assuming a constant velocity), it converged on a spot near the nebula's center, in the year 1140. This imperfect agreement posed the first mystery: Why does the motion of the filaments seem to be accelerating?

In 1949, Australian radio astronomer John Bolton discovered that the Crab is a strong source of radio waves. But unlike what is seen for some other sources of such waves, the radio flux falls off only slowly at higher frequencies, much more slowly than would be expected if the radiation were given off by a hot gas. The slow fall-off meant that total amount of energy involved was surprisingly large. So the second puzzle was where all the radio energy came from and why its radiation didn't have the behavior expected if it came from a hot gas. When our Moon passed in front of the Crab in 1964, astronomers got another shock. At some frequencies, nearly half the radio energy of the Crab was seen to be coming from a dim star near its center. How could such an apparently insignificant star be putting out so much energy? And in 1963 a small rocket that carried an X-ray detector above the atmosphere recorded evidence that the Crab nebula is a powerful source of X-rays. This only made the energy dilemma worse.

In the 1950s, Russian astrophysicist Iosif Shklovskii proposed a solution to the mystery of the Crab's radio energy. He suggested that a phenomenon well known to particle-accelerator physicists dominates the central part of the nebula. As high-energy electrons spiral in a magnetic field, they give off radiation—not only radio waves but a ghostly glow of visible light called *synchrotron emission.* Synchrotron theory required that the waves emitted this way be polarized, that they vibrate up and down in a particular plane. Therefore Shklovskii predicted that the diffuse light of the Crab would be polarized. This would make the Crab look different when viewed through a polarizing filter like the material Polaroid sunglasses are made of. He was right. As seen in photographs made through a polaroid rotated in different orientations, the hazy white cloud of the nebula changes shape dramatically. So the cloud isn't hot gas at all, like all the previous clouds astronomers had seen in space. It is eerie synchrotron light given off by electrons trapped in an intense magnetic field. By contrast, light from the filaments isn't polarized—it comes from glowing hydrogen and oxygen atoms. (A collection of glowing atoms gives off light with random directions of vibration.) Finally, by rotating their antennas, radio astronomers showed that the Crab's *radio* emission is polarized, as it must be according to synchrotron theory.

Shlovskii's ingenious solution of one problem posed another: What energy source sustains the fast-moving electrons racing around the nebula at nearly the speed of light? As electrons radiate, they give up energy. Unless energy was constantly being fed into the system somehow, the electrons would lose their speed and the glow would fade. With the discovery of the Crab's X-rays, the mystery deepened. The total energy required was more than a hundred times that given off by the Sun. As often happens in science, the answer came from an unexpected direction.

During the late 1960s, Jocelyn Bell and Anthony Hewish were working with radiation from quasars, distant and very powerful radio sources. Their antenna covered several acres, but because it wasn't steerable, they had to wait for the Earth's rotation to carry the antenna under each source. On November 28, 1967, Bell noted a signal that would baffle the world of astronomy. A series of equally spaced pulses repeated each 1.337 seconds. After eliminating the possibility of terrestrial interference, the researchers jokingly associated the signal with intelligent beings; they called them *LGMs,* for little green men. Since most phenomena in astronomy involve huge objects, they are associated with long time scales, not short ones. So it was hard to think of astronomical causes for a string of beeps just a second apart. Only after Bell and Hewish found several more sources of regular pulses in different parts of the sky, with different periods, were they sure they'd discovered a new natural phenomenon and not signals from an extraterrestrial civilization.

In a worldwide blitz, radio astronomers found dozens of these new *pulsars,* some blinking considerably faster than once a second. Theorists soon eliminated all but one explanation: Pulsars, they concluded, had to be rotating neutron stars, about 10 kilometers in diameter. Nothing larger could survive the huge acceleration created by such rapid vibrations or rotations. Ordinary stars, even white dwarfs, would be torn apart. Thomas Gold showed how a neutron star formed in the collapse of a massive star would be born spinning rapidly. Most stars rotate. Just as a spinning ice skater increases her speed by bringing in her arms, a collapsing rotating star would leave a fast-spinning remnant. Stars also have magnetic fields. In a collapse, the field strength would increase enormously as the distance between lines of magnetic field shrank—to levels unattainable in magnet labs on Earth. An extremely dense star would also be surrounded by ionized gas, with an ample supply of free electrons. Gold speculated that the spinning magnetic field would gather up these electrons and accelerate them to near the speed of light. The synchrotron effect would then produce beams of radio waves that swept around the neutron star like a lighthouse beam. By chance, these beams illuminate the Earth. This mechanism neatly accounted for the fast, regular radio pulses that had so excited astronomers. Although the details are still in question, Gold's explanation has survived until today.

Soon after, astronomers at the National Radio Astronomy Observatory in Green Bank, West Virginia, found a radio pulsar at the center of the Crab nebula. It was flashing faster than any yet discovered, more than thirty times a second. But which star was the neutron star? Radio telescopes don't have fine enough resolution to tell. Later, astronomers at Kitt Peak National Observatory in Arizona recorded 33 millisecond pulses of ordinary visible light from one of the stars near the center of the Crab—the same star Caltech physicist Fritz Zwicky had suggested might be a neutron star back in 1937. And the star was the one closest to the point that the filamentary expansion pointed back to. As the whirlwind of discovery continued, astronomers at the world's largest radio telescope at Arecibo, Puerto Rico, detected a slight slowing of the Crab radio pulsar. How fast it slowed was consistent with the amount of energy the Crab gives off in all forms of radiation. In other words, the rotational energy of the spinning neutron star is continuously transformed into radiation. The central mystery of the Crab nebula was solved at last.

Whirling at the nebula's heart, the Crab neutron star drives a complex energy machine, producing each form of radiation we detect. Caught up by intense magnetic fields, electrons are flung violently outward, spiraling all the while around field lines and radiating (by synchrotron emission) in every energy band from radio and microwaves to visible light, X-rays, and gamma rays. The high-energy electrons catch up to and push outward on the filaments, which the supernova explosion expelled long before. Their intense pressure, and that of the magnetic field, force the whole filamentary expansion to accelerate. This explains why the expansion seemed to start at 1140 rather than 1054. Thrusting relentlessly outward, the synchrotron nebula also disrupts the filaments into smaller and smaller pieces, giving it the intricate lacy structure we see today.

The spinning magnetic dervish traps more massive charged particles too. It may, some astrophysicsts speculate, accelerate protons and heavier nuclei to energies beyond that achieved at even the most powerful particle accelerators on Earth. These high-energy particles could explain the so-called "cosmic rays," along with other charged particles accelerated in the original supernova blast. These cosmic rays play a surprisingly large role in the energy balance of the universe—their total energy is about the same as that of starlight.

Now that we know about the neutron star at the center of the Crab nebula, we can reconstruct the stellar disaster that led to the nebula's birth. A star not too different from our Sun, but perhaps eight to ten times more massive, once shone at the location of the present neutron star. The star eventually exploded in a supernova, as we described in the previous chapter. At the moment of death, its iron core collapsed, triggering the formation of a rigid neutron inner core. All the rest of the core bounced, blasting charged nuclei, electrons, and radiation outward. The blast forced its way through the outer layers of the star irregularly, breaking much of it up into filaments flying off at speeds of up to 10,000 kilometers per second. Left behind, spinning at perhaps a hundred revolutions per second, was a neutron star of 1.4 solar masses, about 10 kilometers in diameter. No material we know on Earth could possibly survive such violent rotation. Yet a neutron star, basically a giant single nucleus held together by the same strong nuclear force that binds nuclei of ordinary matter, can.

Initially the magnetic field at its surface is about a trillion times stronger than the magnetic field at the surface of the Earth—the field that makes compasses point north. A person wearing a steel belt buckle unfortunate enough to be surrounded by such awesome magnetism would be flung about at supersonic speed. As the newborn neutron star and its magnetic field spin, their ability to accelerate charged particles and generate electromagnetic radiation are at a maximum. Energy is lost at a much faster rate than will be possible later. The spinning slows. Sometimes there are sudden changes in the rotation rate. These "glitches" or neutron "starquakes" may mark sudden changes of shape from more flattened to more spherical, or the expulsion of a large burst of high-energy electrons—an especially large glitch was observed in 1969.

New images taken with the Hubble space telescope reveal more and more intricate detail within the Crab nebula. These high-resolution pictures show entirely new structures and help clarify the chemical composition and temperature variation of each filament. Carbon, oxygen, nitrogen, sulfur, and other elements show up clearly. But with new detail has come new mystery. The relative abundances of certain elements don't seem to match with theory. Far more dust is sprinkled throughout the filaments than was thought possible. Also, there's evidence for more argon than in other supernova remnants, and the argon emission comes from strange little glowing knots lined up with the poles of the pulsars. How the knots formed is unknown. The superior sensitivity of the space telescope even reveals a peculiar glowing doughnut-shaped puff on one side of the pulsar and a curious bright knot of gas close to the pulsar on the opposite side.

Adding the neutron star mass, the filaments' mass, and that of the diffuse nebula totals only three solar masses. At least four or five solar masses seem missing, assuming the parent star was at least eight solar masses—the minimum thought capable of exploding as a Type II supernova. This troubling discrepancy may have been resolved. Paul Murdin of the Royal Observatory, Edinburgh, has detected a huge halo of hydrogen surrounding the Crab nebula. He estimates it totals about four solar masses, just what seemed to be missing. From studies of remnants of shattered stars, like that of the Crab, astronomers are becoming more confident that they understand the basic astrophysics of supernovae—how the matter of a dying star is expelled so that new worlds can be born. But filamentary nebulae like the Crab are in strictly limited supply. Even if another supernova went *bang* close by within our galaxy, perhaps even nearer than the Crab, we'd still have to wait hundreds of years for a new nebula to form. There's another frontier in supernova research, a realm where the supply is virtually unlimited, in the billions upon billions of distant galaxies that form our universe.

CHAPTER FIFTEEN

The Supernova Hunters

At dusk, the telescope dome opens. Relays click and motors whir. The sounds stop. Minutes pass in silence but for an occasional cricket. Then more clicking and whirring. The pattern repeats over and over. A distant glow reveals a sleeping metropolis. There's a road, but no car headlights brighten it. No engine sounds are heard, nor footsteps. There's a door, but no one goes in or out. Just before dawn, the slit narrows, the dome closes.

Far away there is a room with six powerful Sun computer workstations, each the envy of any hacker. No PCs or Macintoshes here, no consumer-style little monitors. In one corner is a tower of disk drives, gigabyte upon gigabyte. At each station sits a scientist, casually dressed, peering intently at a large screen. Over one's shoulder, you recognize a field of galaxies, hundreds of them, all shapes and sizes. Less familar patterns follow—spots, squares, symbols, commands, windows within windows. From the next workstation there's a voice: "I think I've got something. . . ."

Supernovae are rare. Those close enough to study in detail are even rarer. This makes it thrilling to find one, but it also makes them hard to study. We have not been lucky enough to find one going off within our own galaxy since the time of Kepler, four hundred years ago. The bright one found most recently, Supernova 1987A, appeared in the Large Magellanic cloud, a small galaxy about 160,000 light-years away—very close compared with the distances of most galaxies. When we find a very distant supernova, our new-found treasure may consist of but a few squares of varied brightness on a computer screen. In a vast universe containing billions of galaxies, we have clues as to where to look. Any galaxy, spiral, or elliptical can suddenly flare with a concentrated light that soon equals its entire background glow.

To put the problem of searching for supernovae in perspective, let us estimate how many we can expect to find. The answer depends on how many galaxies we are able to watch simultaneously. In a typical galaxy, we expect about one visible supernova each hundred years. Thus if we patrol a hundred galaxies, we will find about one supernova per year—hardly enough to keep a research team busy. If we can afford to patrol twelve hundred galaxies, we will find one supernova a month—much better. It would take a survey covering 5,200 galaxies to produce one supernova a week. And if we'd like one supernova a day to study, we'll need to keep tabs on 36,500 galaxies.

The idea of searching systematically for supernovae dates back to a celebrated 1934 paper by Walter Baade and Fritz Zwicky, in which the term *supernova* itself was coined. CalTech astronomer Baade and physicist Zwicky studied the twenty or so extremely bright novae recorded up to that time. They explained these events as explosive transformations of huge stars into tiny neutron stars, whose existence had just been proposed by Lev Landau. At the time, this was very heady talk. James Chadwick had discovered the neutron itself only recently, in 1932. Baade and Zwicky also speculated that supernovae somehow accelerated charged particles to high energies, thus providing an explanation for the baffling cosmic radiation.

In their long partnership, Baade played the straight man, a traditionally trained and meticulously careful observational astronomer. Zwicky was the idea man. His sensational ideas about supernovae garnered much publicity, but the actual data were too skimpy to convince astronomers. Then as now, physicists invading the field of astronomy were regarded as upstarts, and Zwicky was a notoriously abrasive character to boot. Few astronomers were yet persuaded that supernovae were crucial steps in stellar evolution and therefore worth a major effort to discover in quantity. Nevertheless, Zwicky put together a team consisting of himself, Baade, Milton Humason (who had worked with Edwin Hubble), and Rudolph Minkowski, famed as a spectroscopist. Today those names add up to a dream team of superstar astronomers, but at the time they were relatively young, unrecognized researchers.

At first, Zwicky's search was on such a modest scale that it yielded no results. His equipment, a 3.5-inch camera, coupled to a 12-inch refractor,[1] was smaller than the gear used today by some amateurs. Fortunately for Zwicky and the whole field of astronomy, a telescope of innovative design had just been introduced, one ideally suited for scanning large regions of the sky. With one of the first Schmidt telescopes, an 18-inch built on Mount Wilson, Zwicky and his assistant Dr. J.J. Johnson began finding supernovae in distant galaxies. His method was to compare photographs of galaxies taken at different times, using a binocular microscope, and to look for new objects on the later photograph.

Between 1936 and the end of 1941, Zwicky found another fourteen supernovae, and Johnson four. Oddly, all the ones Zwicky found turned out to be of Type I (without hydrogen in their spectra), while Johnson's were of Type II (with abundant hydrogen). After they found each supernova, Baade made measurements to pin down the light curves, while Minkowski captured spectra with the much more sensitive Mount Wilson 100-inch telescope.

Zwicky and his collaborators eventually discovered some two hundred supernovae, mostly using the new 48-inch (1.2-meter) Schmidt telescope on Mount Wilson. In so doing they opened a whole new field of astronomy. Even as of the mid-1990s, nearly one-third of the more than seven hundred supernovae discovered can be credited to Zwicky and his collaborators. While much has been learned using the original techniques, reliance on photography made the process very laborious and introduced a long delay between the image-making and the recognition that a supernova had taken place. This made it difficult to determine the light curves, whose dramatic rise and fall are a crucial tool in understanding what had happened. Even worse, by the time spectra were captured, there sometimes wasn't enough light to get useful results. Finally, the lack of sensitivity of photographic emulsions compared with modern

1. Telescope sizes are customarily described by the diameter of their main mirror (in the case of reflecting telescopes) or of their objective lens (in the case of refracting telescopes).

techniques meant that the most distant supernova remained hidden. As we shall see, the most distant ones are the most valuable to cosmological research, while the nearby ones are more valuable to the study of how stars die.

Finding supernovae by peering into microscopes at photographs is a grueling business. As early as 1939, Zwicky discussed the possibility of using the infant technology of television for astronomy with Dr. Zworykin of RCA, inventor of some of the first TV tubes. Unfortunately, the electronic-imaging technology of that era had a long way to go before it could relieve the tedium of supernova hunters.

Since Zwicky's time, the process of searching for supernovae, and for much other astronomical quarry, has changed almost beyond recognition. Until the late 1960s, telescopes were guided by hand. Being an observational astronomer often meant spending long cold nights in a cage high above the telescope's main mirror. Heat was out of the question, since it would cause convection currents of air above the mirror, ruining the clarity of the image. Some astronomers enjoyed the rigors of life at a mountaintop observatory; others didn't. With the development of computers and electronic-imaging devices to help guide the telescope, it became possible to automate the more laborious phases of astronomy. Today's astronomers can load a computer with a list of coordinates in the sky for regions to be observed. The computer slews the telescope, locks into a guide star from a digitized catalog, and holds the field in view for a predetermined time. So pervasive is computerized guidance today that even moderately priced amateur telescopes take advantage of it. Thanks to electronic methods of acquiring images, most astronomers and astrophysicists have been freed from the tyranny of photography. Professionals spend much of their time at high-powered computer workstations designing or running sophisticated image-processing software. (They themselves might say they have merely exchanged one tyranny for another.)

The last few decades have seen startling advances in imaging technology. Early television cameras weighed as much as a person and cost more than $100,000 but had such poor sensitivity that daylight or bright studio lighting was required to get any image at all. Today for $400 you can buy a camcorder sensitive enough to shoot videos of the kids indoors in dim light. The same technology that makes home video so convenient has revolutionized astronomy. Its central innovation is the charge coupled device, or CCD. With a CCD camera, it is possible to detect much dimmer astronomical objects. Best of all, the images are recorded in digital form, ideal for manipulation by computers. The power of modern digital image processing can now be brought to bear on refining astronomical data analysis.

Just what is a CCD? Basically it's a chip that detects light. Photons of light hit a semiconductor surface, usually silicon. Electrons from the silicon atoms are released into a mode (a so-called *conduction band*) where they can move freely. There's some similarity between this process and the photoelectric effect in metals discovered by nineteenth-century physicists. Einstein was the first to explain how photons incident on metals knock out electrons; the basic concepts apply to insulators and semiconductors like silicon too. The great advantage of using a nonmetal is that the charges created by incident light don't necessarily drift away immediately, as they would in a conductor. The CCD chip is divided up into thousands or even millions of little squares, called *pixels,* which store charge temporarily. The sensitivity of this process is much higher than what happens in photographic film. For good CCDs, up to 90 percent of the photons actually cause a "count" in one of the pixels. In an astronomical CCD camera, a shutter stays open for the length of the exposure, say 10 minutes. Then the shutter closes and readout begins. It's a kind of electronic bucket brigade, which is where the name *charge coupled* comes from. Charge is moved from one pixel to the next by applying a series of voltage pulses to the electrodes that make up the pixels. Since the timing of the pulses is known, the readout electronics can extract the number of counts in each pixel based on its X and Y coordinates. The number of counts determines the image brightness. Finally, CCD images end up as files of numbers on a computer disk.

Although the electronics involved in reading out CCD chips is more involved than we have intimated, the chips themselves are far simpler than computer microprocessors such as a 486 or Pentium chip. They are somewhat similar to memory chips. Astronomers were fortunate that CCD technology was initially developed by companies such as Fairchild, RCA, and Texas Instruments for video, space, and military surveillance purposes, because the astronomy community would not have been able to foot the bill themselves. Today's latest chips have 2,048-by-2,048 pixels. That's 4,194,304 in all, more than ten times the number found in a camcorder CCD. The resolution of telescopes that use such chips is about 0.5 arc seconds per pixel, enough to see a person on the ground from an orbiting telescope several hundred miles up!

For some astronomers, computer control and electronic imaging were only a beginning. They nurtured the dream of a fully automated robotic observatory, running silently without human intervention. The dream is motivated more by economics and convenience than by fear of frostbite. A robot telescope on a mountaintop could observe flawlessly night after night, while its astronomer masters tended to business in the city.

The struggle to develop automated telescopes to search for supernovae lasted for decades. Until recently, astronomers' visions have run far ahead of available instrumentation. The first successful semiautomated program began in the 1960s, using a specially built 24-inch telescope at Corralitos Observatory in New Mexico. Northwestern University astronomers led by Allen Hynek were able to find fourteen supernovae in relatively nearby galaxies. Their image-recording devices were TV tubes. These devices had evolved greatly since Zwicky's musings in 1939, but their sensitivity and resolution don't compare with that of a modern CCD. (Hynek's process required tedious real-time visual comparison between a TV image of a galaxy and a reference image, since it was difficult at the time to record the images digitally.)

Although he was never able to use his system to discover supernovae, Stirling Colgate of the New Mexico Institute of Mining and Manufacturing designed and built the first fully automated telescope for supernovae. In the late 1960s and early 1970s, Colgate, a leading supernova theorist as well as president of the institute, converted a surplus military radar mount to carry a 30-inch telescope. Colgate wanted to discover supernovae early in their rise to brilliance. Most previously discovered supernovae had been found after their maximum, too late to take the spectra Colgate needed to test his elaborate models for explosions of massive stars. His original concept called for digital data to be beamed by microwave from South Baldy Mountain to the campus nineteen miles away. Unfortunately for Colgate, the age of the microchip had not yet dawned. He was stuck with inefficient TV tubes and room-sized computers less powerful than today's laptops. After twenty years of pioneering effort, his telescope was still not ready to take useful data.

The impact of Comet Shoemaker-Levy 9 fragment G on Jupiter as observed in the infrared on July 18, 1994. The huge fireball at the lower left, larger than the size of Earth, was seen 12 minutes after impact. On the right, the fragment A impact site is also visible. PETER MCGREGOR (AUSTRALIAN NATIONAL OBSERVATORY) USING THE 3-METER TELESCOPE AT SIDING SPRING, AUSTRALIA

The Moon showing extensive cratering. Most of the large craters were formed by huge projectiles billions of years ago, but as on other bodies in the solar system, impact cratering continues at a lesser rate into the present. Were it not for erosion, vegetation, and the existence of the oceans, the surface of Earth would have a similar appearance. The photograph was taken shortly after the Apollo 17 spacecraft left the Moon on its homeward journey to Earth. PHOTOGRAPH COURTESY OF NASA

Meteor crater in Arizona, formed by an impact about 50,000 years ago. Nearly a mile wide, this is the best preserved impact crater on Earth. PHOTOGRAPH COURTESY OF DAVID J. RODDY, METEOR CRATER, NORTHERN ARIZONA

Mead crater on Venus. This radar image taken by the Magellan *spacecraft shows Mead crater, the largest impact crater on Venus. Over 900 impact craters in all were observed on Venus from only a few kilometers in diameter to the 280 kilometer Mead crater. Its multi-ring form is typical of very large craters in the solar system.* PHOTOGRAPH COURTESY OF NASA

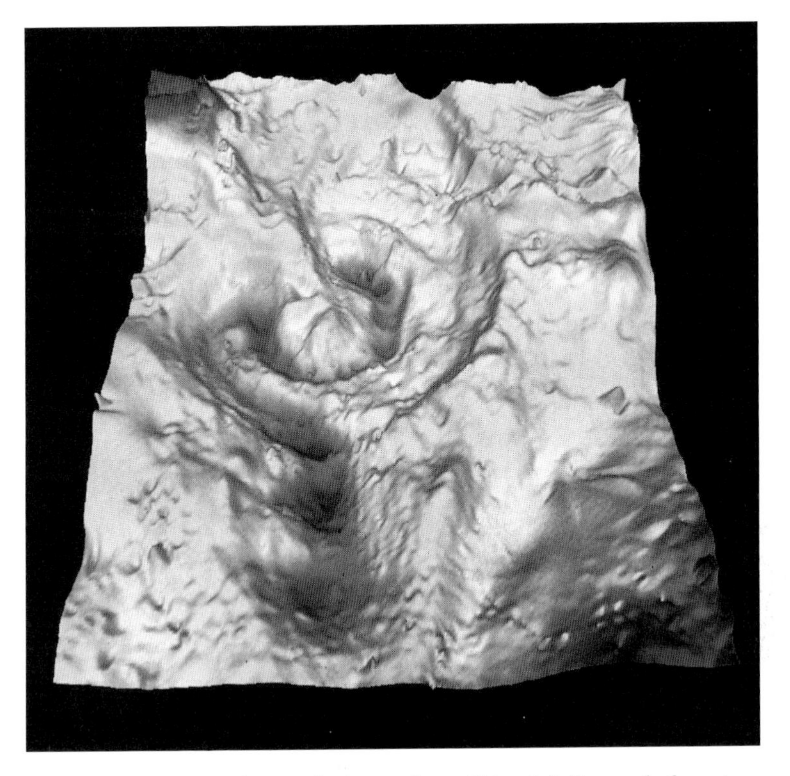

Chicxulub Crater. This gravity image shows Chicxulub Crater, the largest impact crater known on Earth. About 170 kilometers in diameter, the crater is hidden partly beneath the Yucatan peninsula of Mexico and partly under the Caribbean Sea. The crater is revealed by the differing densities of rock within its structure, which produce small differences in gravitational attraction. COURTESY OF DR. VIRGIL L. SHARPTON, CENTER FOR ADVANCED SPACE STUDIES

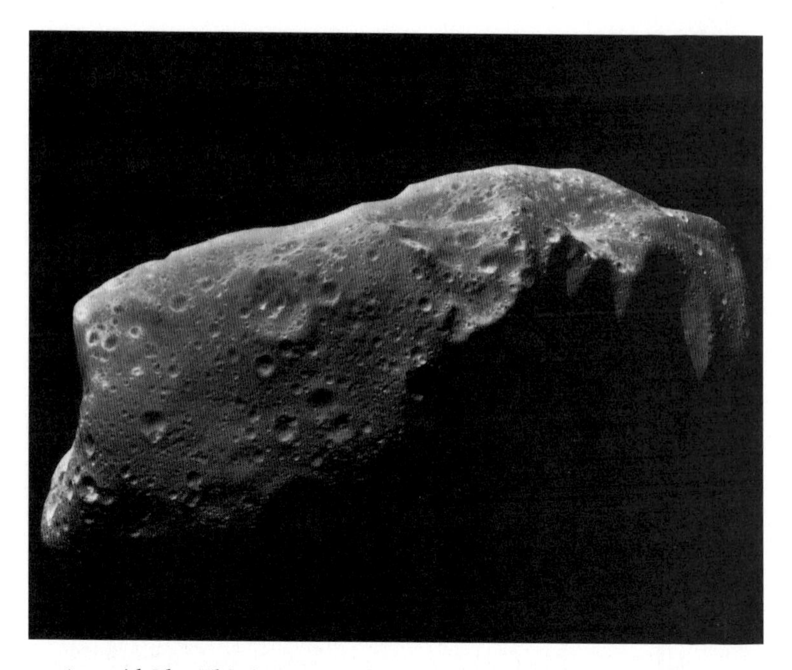

Asteroid Ida. This image acquired by the Galileo *spacecraft from a distance of about 3000 kilometers shows a heavily cratered asteroid about 55 kilometers long. Ida was only the second asteroid to be imaged at close range. Much computer processing was necessary to produce such a sharp picture.* PHOTOGRAPH COURTESY OF NASA

The Nucleus of Comet Halley. This composite photograph taken by the Giotto *spacecraft on March 14, 1986, shows the complex surface of comet Halley's nucleus, which is about 15 kilometers long by 8 kilometers wide. The nucleus rotates with a period of about 54 hours. The sun illuminates the nucleus from the left of the picture. The bright areas are source regions for active dust jets. This object is somewhat larger than that believed to have caused the K-T catastrophe of 65 million years ago.*
PHOTOGRAPH COURTESY OF HAROLD REITSEMA, BALL AEROSPACE

Comet Mrkos, photographed on August 26, 1957. One of the most spectacular comets of recent years, the long, straight part of the tail to the left is composed of ions while the fainter curved part on the right is made of dust. PHOTOGRAPH COURTESY OF PALOMAR/CALTECH

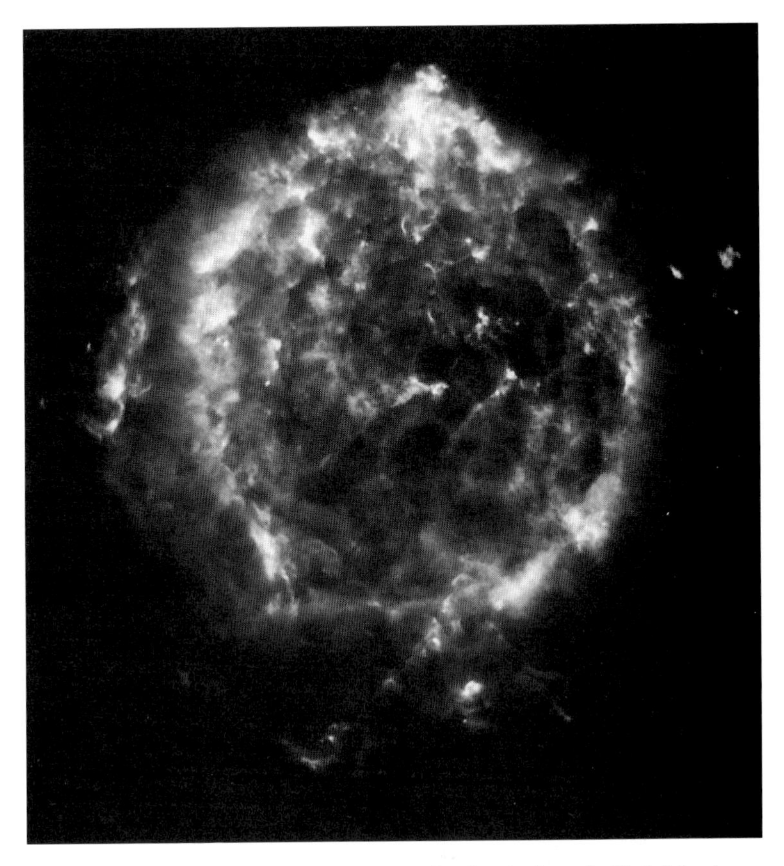

Radio image of the supernova remnant Cassiopeia A, probably left by the supernova of A.D. *1680. Expanding material from deep within the star is breaking through a decelerating shell of material ejected from the star's outer layers, forming conical extensions and crater-like structures between them.* Courtesy of the National Radio Astronomy Observatory, operated by Associated Universities, Inc. Observers were P.E. Angerhofer, R. Braun, S.F. Gull, R.A. Perley, and R.J. Tuffs

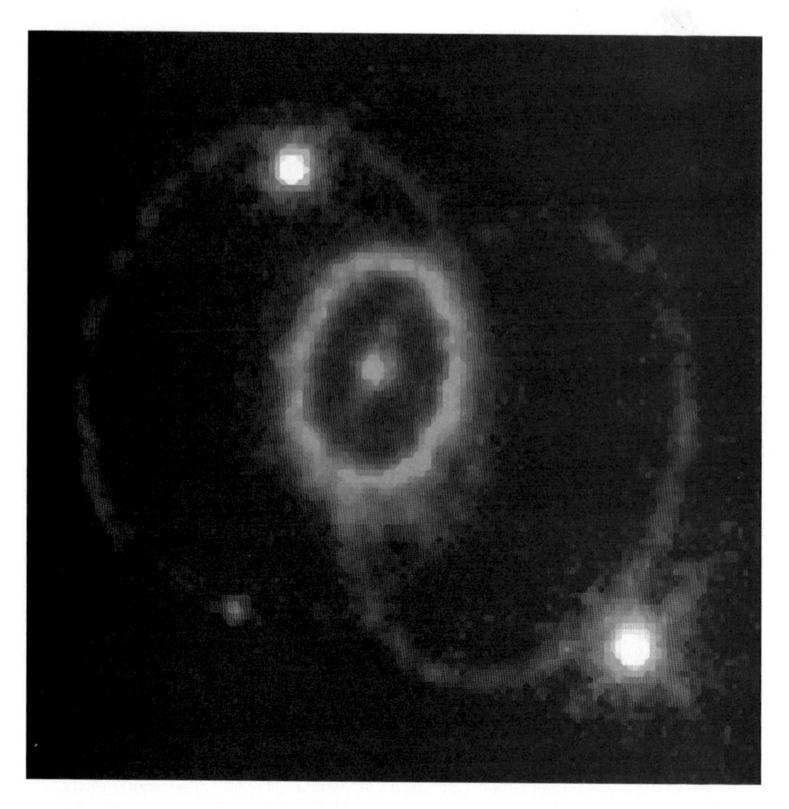

Rings around Supernova 1987A. The rings seen in the space telescope photograph are believed to be formed by light reflecting off clouds of interstellar dust between the supernova and our position. Some astronomers have called them "light echoes." PHOTOGRAPH COURTESY OF NASA

The Crab Nebula in Taurus, source of much current knowledge about supernova explosions and their remnants. The Crab Nebula comprises expanding debris from an explosion witnessed on Earth in the year A.D. *1054.* PHOTOGRAPH COURTESY OF PALOMAR/CALTECH

Deep Sky CCD image of Galaxy Cluster Abell 370. This extraordinary image was taken with the Kitt Peak 4-meter telescope by Don Groom and Saul Perlmutter. It shows a rich cluster containing over 400 separate galaxies located roughly 4 billion light years away. Two supernovae were visible at the time the image was recorded, as indicated by the arrows. The bright curved line near the center of the image is evidence of gravitational lensing, due to the bending of light by the cluster's powerful gravity. PHOTOGRAPH COURTESY OF DON GROOM AND SAUL PERLMUTTER

The Great Galaxy in Andromeda. A great whirling galaxy similar to our own Milky Way, located about 2.3 million light years away. Andromeda is hurtling toward us (or we toward it) at a rate of about 80 kilometers per second. PHOTOGRAPH COURTESY OF PALOMAR/CALTECH

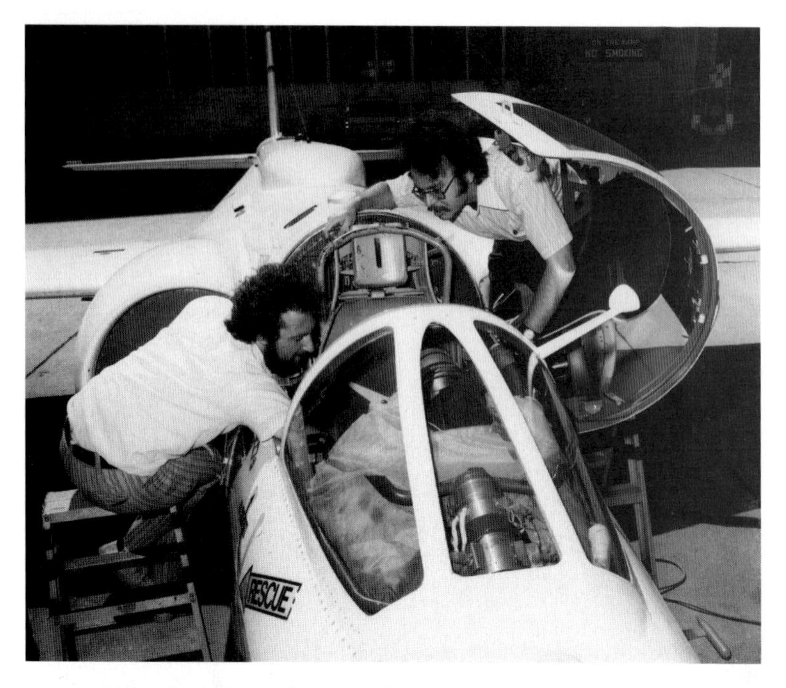

Dr. Richard Muller and Dr. Mark Gorenstein working on the Dicke radiometer in a U-2 aircraft. PHOTOGRAPH COURTESY OF UNIVERSITY OF CALIFORNIA/LAWRENCE BERKELEY LABORATORY

The microwave sky as seen by the Cosmic Background Explorer (COBE) Satellite. This series of images shows the microwave sky after successive stages of background subtraction. The top image shows the so-called "cosine in the sky" due to the earth's motion through space. In the second image this effect has been subtracted revealing irregularities in the primeval microwaves coming to us from the Big Bang as it was only 500,000 years after "the beginning." The horizontal band is due to the emission of our Milky Way galaxy. In the bottom image this excess emission too has been subtracted. The speckles in this image, with a minimum angular size of about ten degrees, are still too large to correspond to any structure yet seen in today's visible universe. PHOTOGRAPH COURTESY OF NASA

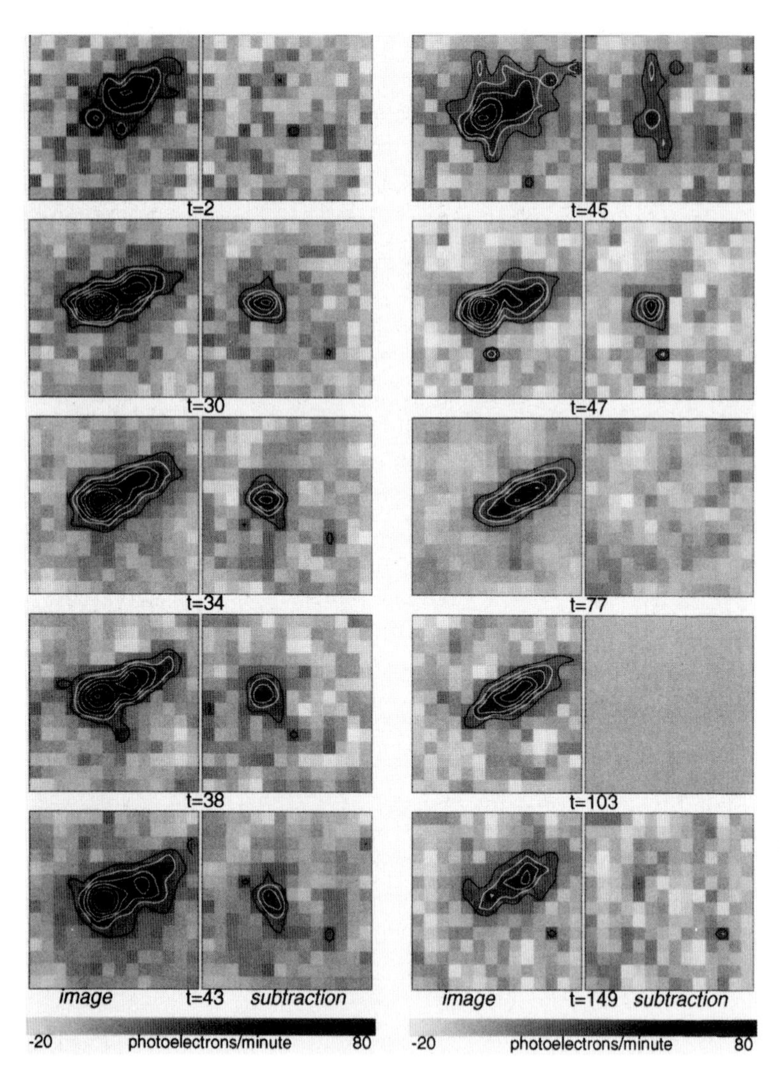

CCD images of supernova 1992bi, the most distant supernova discovered to date. This series of images shows, in each pair, the host galaxy (to the left), and the subtracted supernova (to the right). The images span a period of 149 days from the appearance of the supernova to its disappearance two months later. The supernova and host galaxy have a redshift of z=0.458, corresponding to a distance of about 4 million light years. COURTESY OF UNIVERSITY OF CALIFORNIA/LAWRENCE BERKELEY LABORATORY

Inspired by Colgate, astrophysicists at the University of California Lawrence Berkeley Laboratory renewed the automated supernova search program. Starting just as good CCDs and inexpensive personal computers were becoming available, they didn't have to push the technology so hard. In 1978, Luis Alvarez found out that the air force uses automated telescopes to monitor missile launches. Luis Alvarez and Rich Muller proposed to use air force telescopes on Kwajalein atoll in the Pacific to image galaxies. The proposal was rejected by the air force, but Muller and colleague Carl Pennypacker decided to pursue the project with other telescopes. Within a few years they had an automated search system at the 30-inch Leuschner telescope operated in the Berkeley hills by the University of California's astronomy department.

By early 1986, the group had amassed two thousand digitized reference images of galaxies and were capturing several hundred search images a month. With pictures in digital form, as with photographs, bright supernova are easy to spot. Dimmer flare-ups demand careful pixel-by-pixel subtraction of the reference image from the corresponding search image.

In 1986 the group enjoyed its first triumph with the early detection of a bright supernova in the nearby galaxy M99. Between 1986 and 1990, the Berkeley automated search discovered more than twenty supernovae with their prototype system. Many more such discoveries followed. Theorists were puzzled by a higher-than-expected rate for a type of event that previously had seemed rare. Just as in Zwicky's time, collaborating astronomers trained larger telescopes on the still-brightening supernovae to make precision spectra, which help characterize the supernovae as to type and determine their distance. Groups around the world now proposed to build their own automated telescopes for supernova detection, to find Earth-crossing asteroids, and to search for ever more distant galaxies. Automated astronomy had come of age. By 1990, the Leuschner Observatory was recording galaxy images routinely in the unmanned mode.

It is an eerie experience for an astronomer to stand outside a dome, his or her presence totally superfluous, while a whirring robot telescope goes about the mission of discovery. Between images, the telescope slews from one galaxy field to the next, centers itself on a guide star, then opens the CCD shutter. Just before dawn, the slit closes. If the humidity tops 90 percent, warning of rain, the slit will also close. Each morning a new batch of images arrives in the scanning room via high-speed data-transmission lines.

In order to find supernovae at truly cosmological distances, which could help answer questions about the universe as a whole, the new digital techniques had to be tried on larger telescopes. Using the Isaac Newton 2.5-meter telescope in the Canary Islands, the Berkeley team, now led by young astrophysicists Saul Perlmutter and Carl Pennypacker, detected the most distant supernovae ever seen—in galaxies over 5 billion light-years away. These discoveries may permit a decisive determination of whether the universe is open and fated to expand forever, or closed and doomed to fall back on itself.

The most typical Type I supernovae appear to have the same intrinsic brightness—that is, they all give off the same total amount of energy. This wouldn't be surprising if, as current theory demands, the parent stars were all white dwarf stars of the same mass. Astronomers call objects with that rare and valuable uniformity *standard candles.* They are like a host of candles in a cathedral—the closer ones appear brighter, but since all are identical, it is possible to figure out how far away each is based on how bright they look. We'll return to the story of how supernovae are being used to take the measure of the universe in Chapter 22.

In our story of origins, supernovae play their most brilliant role as prodigal parents. They seed space with the heavy elements necessary to the formation of life and even to the rocky planetesimals from which planets coalesced during the birth of the solar system. Astronomers have long sought a trigger mechanism to explain why molecular clouds, precursors to star formation, begin their collapse, and why some fragment. Some have suggested that shock waves from supernova explosions may have been responsible for the initial compression, allowing gravity to complete the job of cloud collapse. The presence of unexpectedly large amounts of certain radioactive isotopes in some meteorites has seemed to support this contention. However, other astronomers point to less violent explosions as a more likely trigger, maintaining from computer simulations that supernova shock waves are so powerful that they shred molecular clouds rather than squeeze them. Still other astrophysicists think that gravitational instability alone can collapse and fragment clouds, so that we need not seek an external cause for the formation of stars and planets.

Let us put our second big bang, the supernova, aside temporarily and consider the only known bang that is bigger: *The* Big Bang, the one you probably heard about first, the creation of the universe. In investigating this third enigma of violence, we shall find phenomena even more bizarre than the chaotically unstable asteroids and outgassing comets of the first part of this book, and the whirling-dervish neutron stars of the second. Prepare yourself to encounter curved space, the fourth dimension, the Great Attractor, primeval microwaves, antimatter, the X-boson, quarks, inflationary theory, quantum fluctuations, and the ultimate singularity out of which spacetime was born.

III

Creation of the Universe

CHAPTER SIXTEEN

Creation

What exactly is the Big Bang, the one scientists have been talking about for fifty years? Did anything exist before it? Will anything happen after it? Is the Big Bang theory in trouble, as newspaper articles suggest from time to time? Are alternative theories of our origins waiting in the wings, in case Big Bang theorists stumble? These are tough questions, but we'll do our best to start answering them in this chapter.

Most scientists don't like to answer the question of what happened before the Big Bang, because they don't know the answer. Like experts in other fields, they are reluctant to reveal the limits of their knowledge. And yet scientists love to swim in uncharted waters; they surround themselves with mystery and unanswered questions. They love to wonder what questions they should be asking—for the greatest challenge in science is to ask the proper question.

The Big Bang is supposed to have been the first event in the universe, perhaps the one act of divine creation from which everything else has resulted according to the laws of physics. In fact the concept is not quite that grandiose. Most scientists who have seriously studied the subject believe that about 10 or 15 billion years ago, the universe was extremely hot and flying apart from itself at enormous velocities, as if it were exploding. One popular term for the times just after the Big Bang is the *primeval fireball.* Back then, it was too hot for atoms or even nuclei to exist. In fact, the farther back in time you go, the hotter the universe was—thousands of degrees, millions, billions, and even trillions of degrees if you go back far enough. Ever since this hot beginning, it has been expanding and cooling, like steam released from a pressure cooker. The evidence for this is ample, but for now let's just assume it's the only picture consistent with our observations.

How did this state of affairs come to be? The truth is—and this is the answer to the questions about what happened before—we haven't the foggiest idea. Good scientists have made some wild speculations, but just because scientists make them does not necessarily qualify these speculations as science. Not yet, anyway, not until a firmer foundation can be laid beneath them and ways found to test them. But the playful speculations of scientists are at least based on what is known and what could someday be shown consistent with fact. They might better be considered gropings for the right question.

Good science fiction is allowed to violate physical law, but not too often or it loses its authenticity. Likewise, good science speculation must be consistent with what we already know, or must not blatantly contradict well-established physical laws. Remarkably, not even the weirdest discoveries of astrophysics violate the laws that physicists have confirmed here on Earth. In fact, as we shall see, the greatest victories of Big Bang theory have come precisely by applying knowledge gleaned in the laboratory to the alien environment of the early universe. Nevertheless, we must be willing to put aside certain everyday preconceptions when contemplating an event as remote and large-scale as the Big Bang.

The explosion of the universe was profoundly unlike any explosion humans have ever observed. Although it is tempting to imagine this explosion as being *into* something, as fragments of a bomb are hurled into the air, there was nothing for the universe to expand into, and there still isn't. By definition, the universe has no boundaries, no edges; it includes everything that exists. There is no outside.

When the universe was created in a primeval fireball, *it was space itself that exploded,* along with the energy that existed in it. Today space continues to expand steadily in the vast regions between the galaxies. This fantastic notion is far more than coffee-table speculation. It is a natural consequence of general relativity, Einstein's well-tested theory of accelerated motion and gravity, as applied to the evolving universe.

An even more startling conclusion of general relativity is that *space is curved,* and that its curvature is determined by the amount and distribution of masses in space. When objects move or light rays travel, they follow the curvature of space created by nearby masses. Only when the effect of gravity is slight is their path almost a straight line. Yet light and all objects follow the shortest path possible, given the influences acting on the them. In some regions, space is so tightly curved and gravity so intense that light cannot escape; such places are called *black holes.* Although the evidence for black holes is only indirect, and although general relativity is unconfirmed for the scale of the whole universe, the theory has triumphantly passed every experimental test to which it has ever been subjected.

Einstein proposed his *general theory* in 1915, just ten years after he revolutionized physics with his first theory of relativity, called *special relativity.* The special relativity theory, which has now been checked successfully thousands of times, extended Newton's laws of motion to very high velocities, approaching that of light. It demolished the cherished notion of absolute rest, the idea that there is some privileged frame of reference in the universe that is not moving in an absolute sense, and that we could use to measure all velocities. (It is possible to measure the velocity of the Earth relative to the sea of radiation that permeates the universe, but this sea itself is not an immovable, absolute frame of reference, as the terms were understood before Einstein.) The idea that velocity must be measured *relative to* something and that an object can have different velocities depending on what you measure it against was not new. Galileo and Newton understood that kind of relativity hundreds of years before Einstein.

In rethinking the foundations of physics, Einstein built his special theory of relativity on two postulates, which so far have proven to be rock solid. One is that the laws of physics (and the rest of science) are the same in all frames of reference that move with respect to each other at a steady speed and with no change in direction. In other words, there is no way to detect that you are in a fast-moving vehicle, such as an airplane, without looking outside the vehicle. If you doubt this (forget all noise and vibration for the moment), recall what happens when you drop something in a car, train, or airplane; it appears to fall straight down, just as it would if you dropped it at home. Any other "physics experiment" you could perform in the moving and "at home" frames would also give identical results in both places. Einstein's second postulate was that the recently discovered laws of electromagnetism were the same in all reference frames. This assumption has the consequence that the speed of light, 300,000 kilometers per second, is the same in all reference frames. This might not seem shocking until you realize how differently light behaves from, say, a baseball. A baseball thrown swiftly from a truck speeding toward you will move much faster than one thrown by a pitcher standing on the ground—and you'd risk life and limb to try to catch it. Light simply doesn't behave this way—in a sense, its speed is absolute because it is the same to all observers.

From these starting points, Einstein made some bizarre conjectures that far transcend common sense. He didn't just speculate, he produced rigorous mathematical calculations. He treated space and time not as fixed, invariable features of the universe, but as flexible coordinates. His predictions included the shrinkage of fast-moving objects, the slowing down of time at high speed, the possibility of twins aging at different rates (so that one returning from a long space voyage would be younger than the one that stayed home), and the equivalence of mass and energy via the famous formula $E = mc^2$. Furthermore, according to Einstein's relativity, space and time do not exist in isolation as independent concepts. They are connected in such an inseparable way that cosmologists—scientists who study the universe as a whole—often refer to them not separately as space and time but together in one word: *spacetime.*

When we talk about the size of the universe, we usually mean its spatial dimensions, and we think of matter as existing in those dimensions. Most people can visualize this concept easily. But matter exists just as much in the time dimension, even though time is not as visual a concept. In special relativity, Einstein had to dispense with the notions of absolute time and absolute motion. There is no one time that is valid everywhere in the universe; only local times, times measured in a particular frame of reference, have meaning. Time itself flows at different rates depending on the speed of the frame where you measure it; in other words, motion in space affects motion in time. It is the intimate relationship between space and time that leads to the astonishing predictions of relativity such as time dilation, length contraction, and the equivalence of mass and energy.

How many dimensions of space are there? In physics, we usually speak of three; one goes from left to right, a second goes up and down, while a third runs forward and back. These are the familiar Cartesian coordinates of high school geometry. In relativity, time is a fourth and coequal dimension. The TV show *Twilight Zone* used to start with Rod Serling's words "There is a fifth dimension, beyond that which is known to man." His fifth dimension was not time but more of a fictional fourth spatial dimension, one that could be entered under special circumstances and provide shortcuts between distant places or allow travel into the future or past. A similar notion survives in science fiction as the "warp speed" of the *Star Trek* series. Interestingly enough, a fourth spatial dimension is useful when we discuss certain models of the universe in general relativity. We cannot observe this dimension, and it does not explain sci-fi time travel or warp drive. A fourth spatial dimension may or may not really exist, but it is helpful to grasp the concept of a fourth spatial dimension in order to understand what cosmologists mean by the expansion of space.

One way to begin to grasp this concept is through an analogy. Imagine a tiny water wave living on the thin surface of a vast ocean—not a creature separate from the ocean but a ripple that is part of it. (In fact, according to the quantum theory of physics, all particles, including water particles, can in certain circumstances be thought of as waves.) This creature carries an intelligence and an awareness of its world, but its awareness is limited. To it, as to all other water waves, the ocean appears flat except for the ripples. In every direction it looks, it sees a nearly smooth expanse of water stretching to a distant horizon. It has never occurred to it that the ocean has depth, nor can it conceive of such a thing, because water waves exist only on the surface. Its limited concept of the water surface is like human beings' everyday concept of space. Space is what we exist in. Space is the only medium in which particles of our matter can exist. Since our body is made up of particles, we can exist only where space is. Any other kind of existence has no meaning for us.

Now suppose one wave moves very fast in one direction until it eventually comes back to the exact spot on the ocean where it left its slower companion wave. The wave returns to its initial position because its space, like the surface of a balloon, curves back on itself. The wave repeats the journey and times it with a very accurate clock. But this time the trip takes a little longer. The next trip takes longer still. What's going on? Soon one water wave guesses that the ocean must be growing.

They still don't know anything about water under the surface of their ocean. No water wave had ever imagined that the ocean had depth, because all they'd ever seen was other ripples moving about on the surface—in what we'd call two dimensions. Eventually the Einstein of water waves comes along and puts forward a brilliant if controversial theory. It says, "Imagine another spatial dimension that is the radius of our water planet. We all know what a radius is because the circles we sometimes see here on our ocean have a radius. Well, this more complicated shape has a radius, but it also has an extra dimension that we never knew anything about before. This dimension is growing, and that's why our planet seems to be getting bigger." None of the water waves can really visualize the new dimension because they can only move on the surface, and they have no way to verify that there even is a region of water down below. Yet if they could measure the total surface area of their universe, they would find it was growing. "How is it growing?" they wonder. "How can the amount of space—water surface—increase?" For us three-dimensional creatures, it is easy to visualize. But those who live in a two-dimensional world have difficulty.

Human astronomers are now quite certain that the universe we live in is expanding—we'll see why in the next chapter—so we're in a similar situation to that of the water-wave beings. But living in a three-dimensional world, we must try to imagine that we are on the surface of a four-dimensional sphere. (It's actually a sphere only if the universe is limited in extent—finite.) This fourth dimension is not time. It's an entirely unobservable spatial dimension. Try to make a clear picture of it in your mind without visualizing a direction in ordinary three-dimensional space. Most likely, you can't!

In the expansion that underlies the Big Bang, all four dimensions are stretching. It's easy enough to imagine the stretching of the first three spatial dimensions, but the fourth, unobservable dimension, the overall "radius" for the expansion, cannot easily be imagined. It is just not a visualizable concept. Here's an explanation that may help you feel more comfortable about it. Take our familiar three-dimensional world and the geometry that describes it. If we insert this geometry into a four-dimensional space, we find that it has a *center.* All points in our three-dimensional world are the same distance (radius) from this center—a point that does not exist in the three-dimensional world but does exist in the fourth dimension. The equations of general relativity don't absolutely require this fourth dimension, but they take a simpler form when it's introduced.

Physicists have not been able to devise a way to test whether the fourth dimension is real or just a mathematician's artifice. But according to the view of general relativity, the universe expands because the galaxies are carried outward by the expansion of space.

There may be even more than four spatial dimensions. Particle theorists have invented these additional dimensions to explain why particles exist. Some theorists think of space as a giant membrane stretched in ten dimensions. They conceive particles like electrons and protons as vibrations in the membrane. (For other physicists, ten are not enough; they need twenty-six dimensions to explain matter!) Although scientists are unsure what the structure of spacetime really is, the Big Bang models that we'll describe later tie our observations of the universe into a neat package. Cosmologists can frame precisely, and soon may be able to answer, some of our most basic questions.

Is the universe finite or infinite? Will it expand forever or fall back on itself? If it does collapse, will it bounce back or just disappear? If it bounces back, will expansion and contraction continue in endless cycles? How big is the universe now? Is it more or less the same everywhere, or is there important structure? Does it consist mainly of the stars, planets, gas, and radiation we see, or is it mostly some unknown matter or other form of energy? Can humanity survive the collapse and reexpansion of the universe? These questions began to enter the realm of the answerable about one hundred years ago, when astronomers first understood what galaxies are—great whirling disks containing billions of stars mostly like our Sun.

CHAPTER SEVENTEEN

The Fleeing Galaxies

Scattered throughout the sky are numerous fuzzy patches of light. Some are clouds of gas and clusters of dim stars within our Milky Way galaxy. Others are separate galaxies, vast whirling collections of billions of stars. Many of these galaxies are shaped like our Milky Way—disks with several spiral arms. Within the arms are bright regions, giant molecular clouds where stars are born. Some spiral galaxies display prominant bars and rings of unknown origin. Other galaxies are elliptical accumulations of stars without interesting visual features. Still others, irregular in shape, are crossed by vast dust lanes that obscure our view of them.

In 1845, Lord Rosse in Ireland completed what was then the largest telescope in the world, with a mirror nearly six feet across and a tube as long as a six-story building is tall. With this unwieldy instrument he discovered the spiral structure of the galaxy known today as M51, and other galaxies. His drawings of M51, known today as the Whirlpool galaxy, show not only the arms but also a companion galaxy, comparable to the Large Magellanic cloud orbiting our own galaxy. But his telescope, big as it was, lacked sufficient resolution to discern individual stars. Nevertheless, Lord Rosse speculated, as had the German philosopher Immanuel Kant as early as 1755, that the spiral nebulae were "island universes" containing countless stars.

Early in the twentieth century, two large telescopes of high quality were built on Mount Wilson, overlooking Los Angeles. With these new 60- and 100-inch telescopes, astronomers could for the first time distinguish individual stars in the Andromeda nebula, another prominent spiral. But even an astronomer peering through a large telescope has no obvious clues as to how far away Andromeda is. As recently as the early 1920s, some astronomers insisted that all hazy spots like Andromeda were clouds of diffuse gas within our own Milky Way galaxy. New evidence would soon destroy this illusion of closeness and set the stage for Big Bang cosmology.

By 1914, at Harvard University's Lowell Observatory, a young astronomer by the name of Vesto Melvin Slipher had succeeded in photographing the spectra (light broken into colors of the rainbow) of certain nebulae. These nebulae appeared to be moving toward or away from us at velocities far greater than those of stars. The Andromeda galaxy appeared to be moving toward us at about 300 kilometers per second, while most other galactic nebulae were speeding away at velocities up to 2,000 kilometers per second. With such velocities, the nebulae would escape from our galaxy, if they hadn't done so already. This was a strong hint that they were not in our Milky Way galaxy at all.

Slipher had found some lines in his spectra that were shifted toward shorter wavelengths, or positions in the spectrum, and some toward longer wavelengths. What did this mean? A galaxy's light comes from its stars. On its way through the outermost regions of a star, some light is absorbed by atoms of various elements. This reduction of light produces narrow dark lines in the spectrum. Physicists know the normal wavelengths of these lines precisely from observations of the Sun and from laboratory experiments. But in Slipher's spectra, all the lines were shifted by a common factor. This suggested that the star that had emitted them was moving toward or away from us at high speed. In physics, it is well known that waves from a moving source, as well as waves detected by a moving observer, will have their wavelength (and frequency) changed. This phenomenon is known as the *Doppler effect.*

When a car speeds past us, we first hear its horn at a higher-than-normal pitch (shorter wavelength). Then as it goes by, the pitch decreases (longer wavelength). Light waves from a moving source behave like sound waves in this respect. In both cases, a wave given off by a source approaching us seems squeezed, or decreased in length. This happens because more wave crests pass us in a given amount of time than if the source had not been moving. An even clearer example is the way water waves hit a boat more often when it heads into waves than when it heads away from them. You can demonstrate the Doppler effect yourself using a large shallow pan of water. When you tap your finger up and down in the water, you create a symmetrical pattern of circular waves. But if you move your finger across the water at the same time as you tap, the distance between waves—the wavelength—is smaller in the direction of motion and greater in the opposite direction.

When a source of light, such as a star, is receding from us, fewer wave crests pass us per second, and the wavelength measured is longer, or redder (since toward the red end of the visible light spectrum, light waves are longer). In astronomy, we say the light has been *red-shifted.* For visible light, this means shifted toward the red—longer-wavelength—end of the spectrum. When a source is approaching us, the wavelength measured is smaller; it is shifted toward the blue—shorter-wavelength—end of the spectrum. We call this a *blue-shift.* If the source is moving relatively slowly, the shift will be small and the color shift will not be significant. A red-shifted line will not necessarily appear red, and a blue-shifted line will not necessarily look blue. But if the source motion is very fast, the Doppler shift may be so great as to move a given "visible line" all the way out of the visible range and into the neighboring infrared or ultraviolet bands. Since we know the positions of spectral lines very accurately, it is possible to use them to determine the relative velocities of stars, galaxies, and other astronomical objects with impressive accuracy. And the velocity doesn't depend on the line we chose, since all lines are shifted by the same factor.

Although the velocities of galactic nebulae, as determined by their Doppler shifts, seemed too large for them to be contained in our galaxy, some astronomers were unconvinced that they were extragalactic. This controversy was finally settled in 1924. While photographing nebulae at Mount Wilson Observatory, the astronomer Edwin Hubble discovered several dim stars, including one in Andromeda, whose light varies over periods of days. He showed that these stars, called Cepheids, were of a type very valuable for determining distances.

Cepheids are supergiant stars about ten thousand times more luminous than our Sun. Because they are so bright, we can observe them even in distant (but not too distant) galaxies. Suffering from a peculiar instability, they expand and contract in a regular pattern. When they are bigger, they are brighter, and when they are smaller, they are dimmer. Some turn almost all the way off and on in a period of a few days. Others, like the North star, Polaris, merely drop in brightness by a few percent. In 1912, Henrietta Leavitt discovered that the brighter Cepheids have longer periods; looking further, she found a simple relation between a Cepheid's period and its luminosity. By Hubble's time, the distances to some Cepheids within our own galaxy were known. In 1924, he discovered a Cepheid star in the Andromeda galaxy that enabled him to estimate the galaxy's distance. From the period he observed for the new Cepheid in Andromeda and Leavitt's relation, Hubble deduced the star's luminosity (basically, how much energy it gives off). From its observed brightness on Earth, he could then calculate its distance, using the well-known inverse square law for the fall-off of observed brightness. Hubble found that the Andromeda galaxy is far beyond our Milky Way, at least several hundred thousand light-years away. Today, the accepted figure is about 2.3 million light-years, more than twenty times the diameter of our galaxy.

With this discovery, the universe immediately became a much grander place. By the 1920s, astronomers had photographed and charted thousands of galaxies. Nearly all were much dimmer than Andromeda and required the most powerful existing telescopes for their discovery. As huge aggregations of stars like our own Milky Way yet so dim, they must be at correspondingly greater distances than Andromeda. The universe, they realized, must be vast indeed.

Hubble and his chief collaborator, Milton Humason, continued discovering Cepheid variables in nearby galaxies, and like Slipher, they made photographs of the galactic spectra. By 1929, they had determined both the velocities and the distance of several dozen galaxies of up to 6 million light-years away. Their data showed a striking trend. Only a few nearby galaxies seemed to be moving toward us. The rest of the galaxies were fleeing away from us at high speeds, some at more than 1,000 kilometers per second. The faster the galaxy was receding from us, the farther away it was.

By 1931, Hubble and Humason had extended their research to galaxies 100 light-years away, with velocities of nearly 20,000 kilometers per second, about 7 percent of the speed of light. They measured their velocity from the Doppler shift and established their distance based on galaxy brightness. The striking trend held up, and their data clearly fit a straight line of recession velocity versus distance. In other words, the galaxies are fleeing from us at velocities directly proportional to their distance. This startling discovery, so basic to cosmology today, is called Hubble's law.

Hubble's law clearly implies that the universe is expanding, but not necessarily in the sense of general relativity. That is, Hubble's observations were consistent both with the (incorrect) idea of an explosion of matter into empty space and with the now generally accepted idea of an explosion of space itself. Einstein applied the equations of general relativity to the universe as early as 1916. He found, to his dismay, that the equations were incompatible with a static universe. If the stars (or galaxies) were not moving (as Einstein assumed) but were just uniformly filling space, their mutual attraction would soon lead to a collapse. To solve this "problem," Einstein added a "cosmological constant," a repulsion factor, to his equations to make the universe static. Had he had more confidence in the consequences of his own equations and predicted that the size of the universe was changing, he would have made one of the greatest predictions of all time. But he didn't, and so the discovery of the expanding universe belongs solely to Hubble.

The notion of expansion leads us to the conclusion that, barring some inexplicable suspension of a basic physical law, the galaxies and therefore the universe were once smaller. Hubble's discovery of the expanding universe is thus the basic justification for the Big Bang theory. But the expanding universe does *not* imply that every single galaxy is moving away from us. The closest galaxies, such as Andromeda, may be gravitationally bound to our galaxy, or they may be moving at random velocities that have no cosmological significance. In fact, Andromeda is moving *toward* us. However, not a single *very* distant galaxy (that is, one many times farther away than Andromeda) has been discovered that is moving toward us. Since individual galaxies and even clusters of galaxies may be bound to each other by local gravitational forces, we should properly think of large clusters or superclusters as the units caught up in the Hubble expansion.

Hubble's law also means that we are not at the center of an expanding universe, indeed that there is no such a center (except possibly in a fourth spatial dimension). To the contrary, an observer in *any* galaxy would see the other galaxies speeding away and measure the same relationship that we do between their speed of recession and their distance. So crucial is this point to grasping the Big Bang that we will explain it for imaginary universes of one, two, and three dimensions. As we have seen, our universe actually has at least *three* spatial dimensions. (The fourth spatial dimension is useful in understanding models of a closed universe.) So each of the following explanations is merely an analogy, not a discussion of the real world.

One-Dimensional Case. Imagine a string of galaxies attached to a stretchable rubber band (see Figure 17–A). The galaxies are equally spaced, with 1 million light-years (Mly) between them. To an observer in the galaxy at, say, 3 Mly on the rubber band, the galaxies at 2 Mly and 4 Mly appear to be receding at the same rate. The galaxies at 1 Mly and 5 Mly, however, seem to be moving away at twice the rate of the closer galaxies, because the stretching rubber band carries them twice as far in each interval of time. Similarly, the galaxies at 0 Mly and 6 Mly recede at three times the rate of the nearest ones, as called for by Hubble's law. An observer on a galaxy at a different position from 3 Mly will have the same result. Hubble's law leads to the conclusion that the universe looks the same to observers in all galaxies.

Two-Dimensional Case. Imagine two arrays of checkers on an expanding rubber checkerboard, representing the spacing of galaxies at two different times in history (see Figures 17–B, a and b). You can think of the change between the two times as coming from the uniform expansion of space (the checkerboard) between the galaxies, which makes distant galaxies appear to be receding. In the third diagram (Figure 17–B, c) the two arrays are superimposed, with the central galaxy at the same position in each. The arrows show the distance traveled by each galaxy as viewed from the central galaxy. But there's nothing special about lining up the central galaxies. You'd get exactly the same pattern by superimposing the first two diagrams at the common point of any galaxy, not necessarily the central one. You can check this for yourself by tracing Figures 17–B, a and b onto plastic transparencies and trying it.

Three-Dimensional Case. Suppose a loaf of raisin bread is rising in the oven. It has been removed from the pan and is now expanding in all three directions so quickly that all distances will double by the time the bread is removed from the oven (see Figure 17–C). Each raisin is a galaxy. The rate that each raisin moves away from the others is proportional to

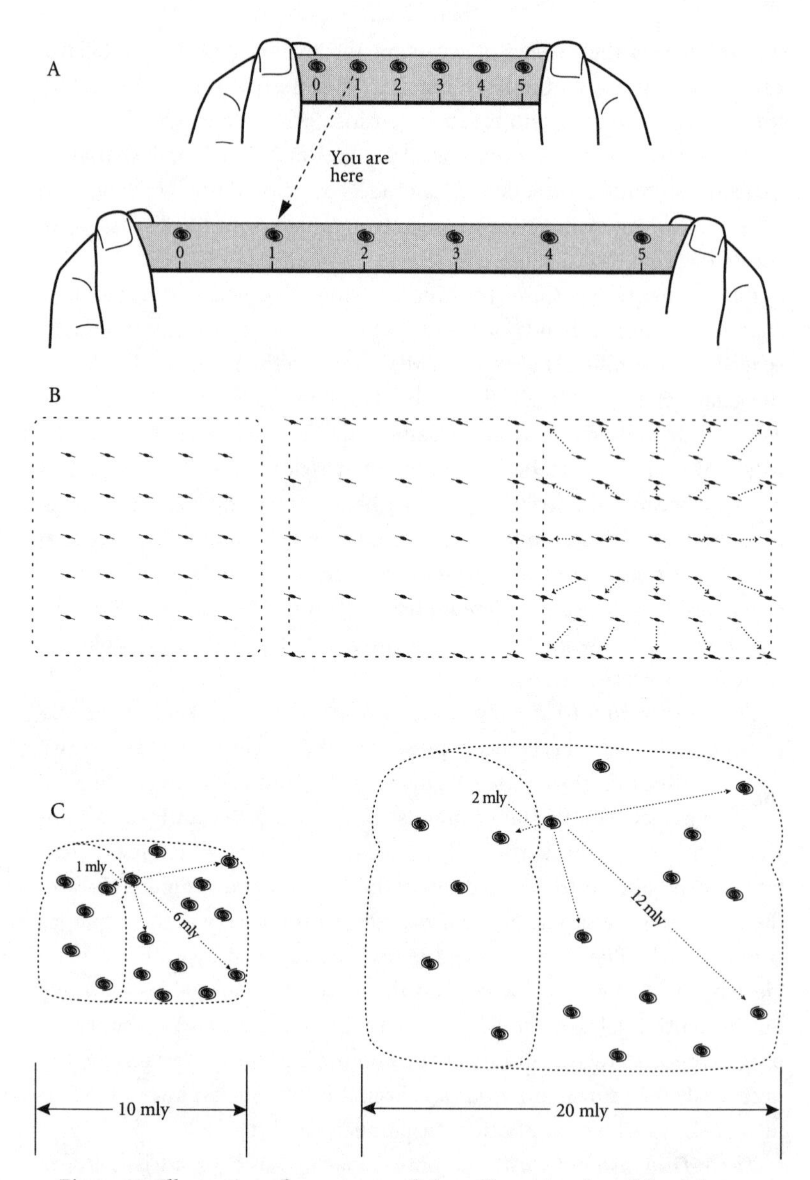

Figure 17. Illustration of one, two, and three dimensional models of the expanding universe showing that an observer in any galaxy would see the other galaxies speeding away and measure the same relationship that we do between their speed of recession and their distance away.

the separation of the raisins. Double the separation, and you double the apparent speed. In the raisin bread, the expansion of the dough drives the raisins apart; in the universe, the expansion of space carries the galaxies to greater and greater distances from one another. An observer on each raisin-galaxy would see all the other raisins receding. Except at the boundaries of the bread, the view is the same from each raisin. Yet the raisin isn't moving with respect to the bread; it is just being carried along.

This analogy breaks down, however, because unlike raisin bread, the Universe has no crust—it is without edges.

Space in all spatial dimensions is being created at a uniform rate. The greater the distance separating two galaxies, the greater the amount of space that is created between them.

This uniform expansion of space applies only to vast intergalactic distances. It does not hold for the distances between massive objects like stars, which according to general relativity strongly influence the geometry of space in their immediate vicinity. Nor does it hold for the distances between molecules and atoms inside matter or between electrons and other subatomic particles. Such distances are controlled by the balance of quantum and electromagnetic forces rather than by gravitational ones.[1] The same is true for distances inside everyday objects and our own bodies; they too are determined primarily by quantum forces. So the Earth will not grow bigger as the universe expands, nor will the bodies of human observers and their meter sticks, which would make the expansion of space and the flight of the galaxies completely unobservable! It is, however, amusing to realize that if we weren't held together by some force, the Hubble expansion *would* make us grow.

The flight of the galaxies is only one of several crucial observations supporting the Big Bang theory, albeit a central one. Each of these observations is elegantly connected to the others. More than anything else, it is the strength of these connections that gives astronomers and physicists

1. By *quantum forces* we mean the equivalent force that derives from the Pauli exclusion principle. The fact that two particles cannot be in the same quantum mechanical state is tantamount to there being a repulsive force at the subatomic level.

confidence that Big Bang cosmology is essentially correct. The "difficulties" and controversies described in newspapers and magazines are in the details—important details perhaps, but the observational bedrock of the Big Bang remains solid. Each connection comes from tracing the universe back to times when it was vastly denser than it is now.

The first such tracing was made by George Gamow, Ralph Alpher, and Robert Herman in the late 1940s. We mentioned it in Chapter 12, in the discussion of the origin of matter. Gamow and his collaborators realized that even if the early compressed universe were made only of hydrogen, other elements would have been created by nuclear fusion reactions. They carried the expansion back to times when the universe was 10^{30} (that's a one followed by thirty zeros, or one million trillion trillion) times denser than it is at present, to the primeval fireball as it existed only a few minutes after the original explosion.

Gamow, Alpher, and Herman assumed the temperature of the universe at that stage to be several hundred million degrees Kelvin—hotter than necessary to start thermonuclear fusion reactions but not so hot that all complex nuclei would immediately break apart into protons and neutrons. Thermonuclear reactions took place at a furious pace during this highly compressed era, fusing some protons, neutrons, and electrons into helium and deuterium (heavy hydrogen) nuclei. But within a few more minutes, expansion of the plasma reduced the temperature and density below the level needed for further reactions. This froze the ratio of helium to hydrogen into a value that scientists think still exists today, some 10 or 15 billion years later. Even though more helium was produced by burning in stars, the amount is negligible compared to that formed in the Big Bang.

Today's universe is a very cold place—virtually all its heat comes from stars. This heat has no more to do with the Big Bang than does the heat from your fireplace. Gamow and his collaborators recognized that the fact of an expanding cold universe could be accommodated by either a hot Big Bang or a cold Big Bang. But only a hot Big Bang could explain the formation of as much helium as we see today. In Big Bang theory, the universe cools as it expands—somewhat as pressurized gas escaping from a spray can cools. Conversely, compression heats, as in a car engine. In a diesel engine the rise of temperature during the compression stroke is so great that combustion takes place as soon as fuel is injected; there's no need for a spark. Thus if the Big Bang ever reverses into a collapsing scenario, we can expect the universe to heat up again.

A hot Big Bang also requires that the universe be filled with radiation. A hot plasma, Gamow, Alpher, and Herman pointed out, would give off and absorb electromagnetic radiation, much as the surface of the Sun emits the light and infrared radiation that warm the Earth. This primordial radiation would be scattered over and over by free electrons until roughly half a million years after the Big Bang. By that time, the density and temperature of matter would have dropped to the point where most electrons and protons would join to form neutral hydrogen atoms. Afterward, the scattering of radiation would virtually cease; in other words, the universe would become clear to electromagnetic radiation, rather than opaque. Whatever radiation existed half a million years from the beginning would be preserved through the vast expansion that followed, although its wavelength would be greatly Doppler-shifted—to longer wavelengths and much lower temperatures.

Gamow and his collaborators predicted the remnant radiation would be faint, with a characteristic "temperature" of about 10 degrees Kelvin—the equivalent of a low-power microwave emission. There was, however, no rush to discover this background radiation, since the technology needed to detect low-power microwaves did not yet exist. As a result, Gamow's prediction of a cosmic background was almost forgotten.

CHAPTER EIGHTEEN

Microwaves in the Sky

Before electronic circuits were introduced to quiet radios and stereo receivers, everyone knew what "static" was—that annoying sound between radio stations. Electrical disturbances in the atmosphere still interrupt broadcasts occasionally and produce this irritating radio noise. In 1965, radio astronomers Arno Penzias and Robert Wilson of Bell Telephone Laboratory were measuring galactic radio noise that could interfere with satellite communications. They pointed their antenna, or radio "telescope," away from the disk of the Milky Way, toward the galactic halo, and picked up a small, strange, background signal that they couldn't get rid of.

Radio telescopes had proven their worth in detecting energy sources in the sky that were hard to see with optical instruments. Astronomers knew that some galaxies give off intense electromagnetic radiation in the radio and microwave bands. Supernova remnants and regions of star formation in our own galaxy were known to be strong sources. Radio waves from the dust-shrouded disk of our Milky Way galaxy helped define its structure by clearly revealing the spiral arms. And in just three years the first baffling radio "pulsar" would startle the scientific world.

But Penzias and Wilson weren't trying to make an astronomical discovery; they were trying to eliminate an apparently spurious signal. Inside their antenna, they found pigeon droppings, which could be radiating. After a thorough cleaning, the signal diminished, but only slightly. Since all electrical circuits produce their own radio noise, Penzias and Wilson's signal might have been coming from their amplifier. But even after they subtracted amplifier noise, as well as radio noise produced in the atmosphere, static remained. The signal, they concluded, must be coming from space.

But the strength of this signal did not depend on the direction the antenna was pointed, or on the season, or on the time of day. If it came from space, it was not being emitted by a single localized object. It was no stronger coming from the direction of the galactic disk. So they concluded that it came not from within our galaxy but from some unknown source.

We should not underestimate their courage in claiming that the signal they were receiving came from outside our galaxy. It was a signal unlike any ever picked up before. It seemed to come from everywhere. Usually, when a signal is independent of the pointing of an antenna, it comes from inside the antenna. To rule out that possibility, you must truly understand your antenna, as Penzias and Wilson did.

The intensity of radiation they measured was what, according to electromagnetic theory, should be given off by a box of matter with walls at a temperature of 3 degrees Kelvin—that is, 3 Celsius degrees above absolute zero. (In modern usage, we say the temperature was "three Kelvins.") Later observers found that the mysterious radiation had a spectrum very close to that expected from a "black body" or perfect emitter of radiation.[1] It is the most ancient signal ever detected, a faint "background" of microwaves coming from behind everything else that astronomers see.

Penzias and Wilson had originally reported that the intensity of the radiation in different directions was uniform to better than 10 percent; measurements by many others soon reduced the figure to less than 1 percent. This feature of the radiation would prove to be most difficult to explain.

Penzias and Wilson stopped short of claiming that what they had found had cosmological significance. They gave their discovery paper the

1. About a hundred years ago, Max Planck initiated quantum physics by deriving the formula for the intensity of radiation (as a function of wavelength) given off by an opaque body at any temperature. Such a body does not need to be black. Any object in thermal equilibrium with its surroundings will do, such as a tabletop, chair leg, or the back of your neck. Each of these everyday objects radiates electromagnetic energy, mostly in the form of heat, or infrared radiation.

modest title, "A Measurement of Excess Antenna Temperature at 4080 Mc/s." But in a paper published at the same time, Princeton astrophysicists Robert Dicke, P. G. Roll, P.J.E. Peebles, and David Wilkinson proposed that the background radiation detected by the Bell Labs antenna was a relic of the Big Bang. Just as Penzias and Wilson made their discovery, Dicke and his colleagues were building their own receiver specifically to detect cosmic microwaves. And Peebles was about to publish a new calculation of the expected temperature. Their argument was similar to that of Gamow and his associates, referred to in the last chapter.

According to simple Big Bang theory, the different parts of the sky that produced the microwaves had not been close enough to each other to come to the same temperature. The only way the intensity could be uniform to even 10 percent was to *postulate* that it had been uniform from the beginning. (Much later, a version of Big Bang theory called the inflationary model resolved this problem.)

In the years following, observers with ground-based radio antennae sought eagerly to detect some directionality to the 3-degree black body radiation, but without success. Calculations of the expected temperature converged on the measured value. Also, only minor departures from a black body spectrum were observed, whereas radio emission from our galaxy has an entirely different spectrum. By the mid-1970s, nearly all astrophysicists agreed that the pervasive microwave radiation discovered by Penzias and Wilson was left over from the creation of the universe, that it was an "echo" of the Big Bang. The Bell Labs researchers received the Nobel prize.

Why is their discovery so important? The existence of this uniform background radiation is, along with the Hubble expansion, the strongest evidence we have for the Big Bang theory. It confirms Gamow's assumption that the early universe was very hot, since the microwaves we observe now would have originally been emitted from a plasma at a temperature of thousands of degrees. The extreme uniformity of the microwave signal tells us that although the Big Bang was unimaginably violent, it happened in a smooth way—much as the surface of the violent Sun is intensely bright but uniform. The cosmic microwave signal emanates directly from the early universe, as it existed a mere half million years after the initial explosion. Signals from even the most distant galaxies and quasars known are younger and come from closer distances. The connection between age and distance is simple; since the cosmic background radiation we observe now is roughly as old as the universe, say 12 billion years, it has traveled a distance of 12 billion light-years at the speed of light to reach us.

Astrophysicists realized early on that study of the microwave background radiation could provide vital clues to the overall structure of the universe and perhaps to the mysterious origin of galaxies—if some directional pattern could be detected. Their quest eventually paid off, though it took more than twenty-five years.

Without directionality, the background radiation seemed to provide evidence that on a large enough scale, the universe is the same in all directions. This uniformity surprised many astronomers because they expected to see bright spots (bright in radio intensity, that is) in the sky where galaxies had formed. They also expected to see some variation in radiation intensity due to the orbit of the Earth around the Sun, and the motion of the solar system through space, carried by the rotation of the galaxy.

The background radiation had an observed temperature of a frigid three degrees Kelvin—yet when matter "froze" into hydrogen atoms from detached electrons and protons (plasma), the temperature was much hotter, about 5,000 degrees. At that instant, the radiation scattered from electrons for the last time, and the matter it bounced off formed a huge expanding shell that completely surrounded our present position in space. Radiation with a characteristic temperature of about 5,000 degrees is visible light and infrared, much like sunlight. Such radiation is what an observer moving with the plasma would see. As a result of the recessional velocity of the plasma from the Earth (due to the expanding universe), however, the radiation is red-shifted from the visible into the microwave region, with a characteristic temperature of about 3 degrees Kelvin. The huge red shift (corresponding to a 1,500-fold increase in wavelength) comes from the enormously high speed of the expanding shell of plasma, as seen from our frame of reference.

Since we prefer to describe the Big Bang as an explosion of space, it's more appropriate to talk about the rate at which *space between us and the shell is increasing* rather than the rate of the expanding shell. As we have seen, the rate of creation of space between us and an object, like the velocity of a fleeing galaxy, is proportional to the distance of the object under observation. As time passes, the waves of cosmic background radiation we observe will have come from even more distant regions of space. Since those regions are moving away from us at still higher velocities, the radiation we observe will be red-shifted even more and therefore will have a temperature lower than 3 degrees Kelvin. When the universe is twice as old as it is today, any astrophysicists left in our galaxy will measure a background temperature of about 1.5 degrees Kelvin, half the current temperature.

After the discovery of cosmic background radiation, astronomers took the Big Bang theory much more seriously. The theory predicted that the spectrum of the radiation (the strength of the radiation at different wavelengths) would be similar to that of a black body, and this similarity is exactly what was observed. Rival theories, such as the steady-state theory of continuous matter creation between galaxies, lost adherents at an accelerating rate. By the late 1970s, the Big Bang had become the standard model for the early universe, and the steady state was almost forgotten. There seemed little doubt that our universe was born in cataclysm and was still bathed in a radiant relic from its birth.

Although most astrophysicists had come to believe that the microwave background filled the universe, they remained anxious about its uniformity. The Earth moves in space; it ought to be moving with respect to the background radiation, and that motion should be detectable as an increase in the intensity and temperature of the radiation in the direction of the Earth's motion—what in technical terms is called an *anisotropy,* (pronounced "an-i-*sot*-ropy"). If this variation could not be detected, something would be seriously wrong—maybe our whole way of thinking about the Big Bang.

Revealing Earth's motion depends on the Doppler effect, the same bit of physics that was so crucial to showing that the universe is expanding. But in this case, instead of merely measuring light received from receding stars, we ourselves are moving through a sea of microwave radiation. An everyday analogy would be the difference in the sound of a honking horn you hear while stationary and in a moving car. When the horn is in the moving car and you're standing on the side of the road, you hear a rise, then fall in pitch. When you are in a moving car and the horn is in a parked car, you hear the same change. As the Earth moves through the cosmic microwaves, similarly, scientists expect to see a brightening of light or increase in temperature in a certain direction.

How fast *is* the Earth moving? It hurtles around the Sun at an average speed of about 30 kilometers per second, but that's just the beginning. The solar system, carrying Earth along with it, is thought to be rotating around the center of our galaxy even faster, at about 300 kilometers per second. And our galaxy is known by Doppler-shift measurements to be moving toward the Andromeda galaxy at about 80 kilometers per second (although we can just as well say that Andromeda is moving toward us at that velocity). Since the velocity of light is about a thousand times greater than the fastest of these speeds, you'd need to be able to see an anisotropy of less than one part in a thousand to measure the Earth's motion. Such a difficult experiment was done, and it had a surprising result. But before describing it, let's take a brief look at a famous earlier experiment intended to measure the Earth's motion.

Back in the nineteenth century, before Einstein had presented his relativity theories, physicists believed that an all-pervading medium carried light waves. Sound waves, scientists knew, could travel only in air or in some other medium, not in a vacuum. Similarly, they thought that light could not travel in the void of space, say between Earth and the Sun, without a medium. This medium was called the *luminiferous aether* (and has no relation to the chemicals known as ethers, other than their shared name). Without any real evidence, scientists postulated that this mysterious substance filled all of interstellar space. Light waves would exist in the aether, just as sound waves exist in air, and water waves exist in water.

From the shore, a water wave appears to travel faster when it is moving with a current than when it is moving against it. Similarly, a burst of wind affects the apparent speed of sound traveling with or against the wind. By analogy, physicists of the 1880s reasoned that they could detect the 30-kilometer-per-second motion of the Earth through the aether by virtue of a supposed aether "wind" created when the Earth moved through the aether. If they had known about galactic rotation, which they didn't, they would have expected a much bigger aether wind.

To detect the aether wind, it seemed reasonable to try to measure the time difference between a trip taken by light in the direction of Earth's motion and a trip taken perpendicular to that motion. To measure the time difference, Albert Michelson, the first American Nobel prize winner, developed an ingenious "interferometer" that had more than enough sensitivity to detect a 30-kilometer-per-second aether wind. But to Michelson's astonishment, he could find no such effect. Try as they might, he and other physicists of the day were unable to come up with a satisfactory explanation.

Only after Einstein proposed his special relativity theory in 1905 were physicists able to understand why Michelson couldn't detect the aether. Einstein postulated a strange new characteristic of time—that it depended on the velocity of the observer. A direct result of this postulate was that the speed of light is the same for all observers, even if the source of light or the observer is moving at high speed. Since the speed of light is constant, there would have been no aether wind even if the aether existed. This postulate accounted for Michaelson's result, but it took many years for physicists to accept Einstein's theory. Einstein was awarded the Nobel prize not for relativity theory but for his more readily understood theories of Brownian motion and the photoelectric effect. Relativity, as we have seen, led to a slew of strange counterintuitive consequences such as time dilation (stretching), length contraction, and the astonishing notion that mass is just a form of energy. But now that special relativity has passed countless stringent tests, it is as credible a cornerstone of physics as were Newton's laws before it.

Against this background, it is easy to see why Princeton physicist Jim Peebles coined the term *new aether drift* to describe the expected motion of the Earth with respect to the cosmic background radiation. But why could a new aether-drift experiment be expected to work where the old one could not? The difference is that the medium being detected in the present case, the microwave background radiation, does not carry light—it *is* light. Special relativity does not contradict the Doppler effect for light measured from a moving receiver, although it does change the calculation of the effect.

Measurement of the new aether drift required instruments that were flown above most of the atmosphere rather than operated from the ground. The reason is that the observations had to be made at shorter wavelengths than in most previous cosmic background experiments in order to avoid interference from microwave radiation emitted by our galaxy. But at these shorter wavelengths, oxygen and water vapor in the atmosphere emit microwaves too. The measurement is possible only above 50,000 feet, where water vapor is frozen out. (Today, experiments can also be performed at the South Pole, where low temperatures achieve the same effect.) Paul Henry of Princeton University, using an instrument carried aloft by a balloon, reported the first result: a small anisotropy in the cosmic background radiation. But his data showed large unexplained fluctuations; although they later proved to be nonexistent, most scientists at the time felt that they could not trust a conclusion based on data so poorly understood. Aside from this problem, the radiation seemed as of the mid-1970s to be uniform to one part in five hundred, thanks to the careful measurements of David Wilkinson and Robert B. Partridge of Princeton and Edward K. Conklin of Stanford University.

To clear up this situation and improve on the previous measurements, Rich Muller began a project at Berkeley in 1976. He was soon joined by a young physicist named George Smoot and a graduate student, Marc Gorenstein. Within a few years, they found the first strong evidence for an anisotropy using an instrument crammed onboard a former U-2 spy aircraft. The device achieved extreme sensitivity by rotating its two-horn antenna (called a Dicke radiometer) once a minute, as well as switching the receiver back and forth between the horns to detect differences in temperature between one direction in the sky and another. A second radiometer detected any asymmetry of the potentially troublesome signal from atmospheric oxygen.

The U-2 experiment gave results that were reassuring in some ways, fascinating and unexpected in others. It was encouraging in that it showed strong evidence of Earth's motion relative to the cosmic background radiation. In one area of the sky, microwaves appeared blue-shifted; indicating the direction toward which Earth is moving. In the opposite direction, from which the Earth came in its overall motion, the data showed a red-shift, as expected. The size of the wavelegth shifts were equivalent to an increased temperature of only one-tenth of a percent—but that was enough to imply a velocity for the Earth that is one-tenth of a percent of the velocity of light! Their best calculation showed the Earth's velocity to be about 400 kilometers per second.

At 400 kilometers per second, Earth's apparent velocity was faster than expected and moving in a direction quite different from that predicted by galactic rotation. Earth seemed to be moving toward a point 15 degrees east-southeast of the star Regulus. So different was this direction from that required by Earth's rotation around the center of the Milky Way galaxy that our galaxy must itself be moving even more rapidly with respect to the cosmic background, the Berkeley team realized. Using vector algebra, they found a value for the galactic velocity of 600 kilometers per second, or more than a million miles per hour. This is the velocity at which the Milky Way galaxy is drifting through the radiation "aether" left over from the Big Bang.

Relative to Andromeda and other local galaxies, however, the Milky Way's velocity is much slower than 600 kilometers per second. So the group concluded that these galaxies must be moving too, as must the nearest large cluster of galaxies, the Virgo cluster. Imagine the situation: We are in a vast volume of space, tens of millions of light-years across, in which are embedded thousands of galaxies, racing at the huge velocity of roughly 600 kilometers per second with respect to the distant universe.

Where did a velocity so fast come from? The present velocity of the Milky Way could be due to local turbulence, but it is difficult to reconcile that possibility with the smoothness of the microwave radiation. The results of the Berkeley U-2 experiment also set a limit on two large-scale properties of the universe: One, if the universe is rotating as some physicists believe, its rate of rotation must be less than one-billionth of a second of arc per century. Two, its expansion (apart from the motion of the Earth, the solar system, and our galaxy) must be uniform to one part in three thousand.

Another balloon-borne Princeton experiment, this one by Dave Wilkinson and Brian Corey, found a cosmic background anisotropy of a size and direction agreeing with the Berkeley U-2 results. These successful experiments gave astrophysicists further confidence in the cosmological origin of the background radiation and stimulated proposals to measure the radiation with much greater accuracy on orbiting spacecraft. The microwave background radiation provides a way to image the universe as it was very early in the Big Bang—possibly the only way we will ever have.

CHAPTER NINETEEN

A Snapshot of Creation

The experiments on the Berkeley U-2 flights made the temperature of the background radiation left over from the Big Bang look smooth in all directions to less than one part in ten thousand (with the exception of patterns of only local significance). Such temperature uniformity posed a tough theoretical problem for astronomers: It suggested that the early universe had homogenized itself in temperature. We can see this process in ordinary matter here on Earth: Even when matter is heated irregularly, it eventually reaches a uniform temperature throughout by various means—conduction, convection, and radiation. But these processes require time: The various regions of the matter must be near enough to one another that light and other electromagnetic waves have enough time to cross from one side to the other. The universe as described by the simple Big Bang theory, however, violates this condition, since regions we can see today are too far apart for light to have been able to cross between them in the age of the universe. There was insufficient time for the universe to attain thermal equilibrium.

The maximum distance that light can have traveled since the beginning of the universe is called the *horizon distance.* Parts of the universe that are now separated by distances greater than the horizon distance can never exchange information like light because no physical process that carries energy can travel faster than the speed of light. They cannot have come to the same temperature because heat cannot have traveled between them. Even though *we* on Earth may be able to see many of these regions, they are beyond each other's horizon.

Parts of the sky that are separated by many times the horizon distance were able to get that way because, according to general relativity, *space can expand faster than the speed of light.* Yet these regions have virtually the same temperature. Simple Big Bang theory doesn't tell us why the temperature should be uniform.

An ingenious addition to the Big Bang model provided a way out of the dilemma. Alan Guth recognized that developments in the theory of elementary particles had unexpected and fascinating implications for the behavior of the early universe. According to these theories, the "vacuum" of space can undergo dramatic changes analogous to the phase transition experienced when ice melts. If we are now living in a frozen vacuum, the early universe was hot and "melted." The mathematics of the new theory showed that in that early period all particles would be massless, like photons, and that they would have acquired mass only when the universe froze. The name assigned to the early vacuum was *false vacuum,* to emphasize how it differs from the present one.

During the freezing period, Guth recognized, the vacuum would behave like a "negative pressure," which would cause an extremely rapid expansion of space. In fact, the expansion would be much greater than the speed of light. But since it is not the objects in space that are moving but space itself that is expanding, Guth called this brief period the *inflationary* phase, and the theory was dubbed "the inflationary universe." The inflation ended at about 10^{-32} seconds after the Big Bang—very early indeed!

The inflationary universe model gave astrophysicists a way out of the horizon-distance dilemma. Before inflation, the universe was small enough that all parts were within each other's horizon distances. There would have been enough time for all the parts to swap heat and come to the same temperature.

Although the possibility of inflation helped account for the smoothness of the microwave background, astrophysicists were still determined to find evidence of cosmological lumpiness. The reason is that the universe we observe today is highly nonuniform; most of the known mass in it is clumped into stars, galaxies, and clusters of galaxies. Nearly all astrophysicists believe that the clumping came about from the mutual gravitational attraction of matter created in the Big Bang. They needed evidence for some structure in the early universe—be it clumping or a variation in temperature—to explain the evolution of the universe into galaxies and clusters of galaxies. A uniform universe without clumping would offer no explanation for galaxies. But this clumping must have begun at a very early stage of the universe and must already have been present at the time when the microwave background began its journey. Therefore, the background radiation we observe today must be nonuniform.

Why must clumping have begun so early? The basic idea is simple enough. By pure chance, some atoms in a cloud of gas will group together in tiny regions of greater density. These regions or "lumps" will pull nearby atoms in because their gravitational attraction is greater than that of the surrounding low-density regions. As the original lumps grow, their attraction for nearby matter increases. Gradually the cloud will separate into large clumps or even collapse into one clump. This tendency is called *gravitational instability.*

In cosmology, gravitational clumping has two powerful countervailing forces. One is the Hubble expansion itself: At first, the expansion was too fast for clumping to proceed, and it hurled matter outward before gravity could pull it in. The other factor opposing clumping is radiation pressure. Until about half a million years after the Big Bang began, so much radiation was emitted that the net effect of forces acting between nearby particles could only have been expansion. Again, clumping would have been impossible. After the decoupling time, when electrons and nuclei "froze" into atoms, the level of radiation pressure plummeted because radiation interacts much less with neutral atoms than with charged particles. But the Hubble expansion remained and was more than enough to prevent gravitational clumping.

Under these conditions, galaxy formation would seem to be impossible. Yet galaxies do exist. Density fluctuations, astrophysicists believe, must have existed from the very early era in the universe when matter first appeared. Again the inflationary model may provide a way out of the dilemma. Various versions of inflation, all rather speculative, allow for the formation of stable structures, possibly associated with the decay of ultramassive particles into the particles we know today, by about 10^{-35} seconds after the Big Bang. Inflation also tends to magnify any preexisting irregularities. Clumps in the range of 10^5 solar masses (that is, clumps with a hundred thousand times the mass of our Sun), and those greater than 10^{12} solar masses could well have survived radiation pressure. (Interestingly and perhaps not coincidentally, these masses correspond to those observed for globular clusters of stars (in our galaxy) in the former case and large galaxies or clusters in the latter.) It also may be that while expansion slows the process of growth for density fluctuations, it tends to stabilize them once formed. That is, it opposes complete gravitational collapse.

Astrophysicists wanted very much to confirm the existence of clumping in the early universe, to account for the formation of galaxies. Despite its successes, the Big Bang theory might not survive if it proved inconsistent with galaxy formation. Finding cosmological anisotropy in the background radiation would be strong evidence of early clumping, so the stakes were high in searching for such anisotropy.

They looked for such irregularities—cosmic in origin—in the microwave background radiation. The U-2 experiments had detected patterns in the radiation, but they were only local in origin, not cosmological. A more precise experiment was needed, and it looked like a satellite would be necessary to carry the apparatus above the Earth's atmosphere. So George Smoot proposed to NASA a project that would use apparatus essentially identical in design to the U-2 project but that would fly in a satellite. It should take two years to complete and launch the apparatus, he said, and another year to take all the data.

Thirteen years later, the apparatus finally flew as part of the COBE (Cosmic Background Explorer) satellite. The long delay was due not to scientific problems but mostly to bureaucratic and political ones, and to bad luck. When NASA needed science projects to justify its space shuttle missions, the project was switched to the shuttle. But it was too small for the shuttle, so it was combined with other formerly separate projects, including ones where humans would be along. But with humans present the safety standards skyrocketed, and so did the cost. Finally, after the *Challenger* explosion in 1986, the project was switched to another satellite.

Launched finally from the space shuttle by NASA in 1989, the COBE satellite carried three radiometers for measuring the microwave background directionality at three different wavelengths. In addition, it carried a "spectrophotometer" (for which John Mather, a former Berkeley student, was chief scientist) to measure the black body spectrum down to far-infrared wavelengths of 0.1 millimeters. Surprisingly, the spectrophotometer measured, to better than 1 percent accuracy, a spectrum with just the shape expected for a radiating black body. Despite some earlier confusion, the astonishing agreement of the COBE data with black body emission was a remarkable confirmation of Big Bang theory.

It was also a tantalizing new puzzle. The spectrum found was that of a black body with a temperature of 2.74 degrees above absolute zero. However it was much closer to that of a black body than had been predicted. The close match is exciting and baffling to cosmologists and it sets strong limits on the nature of material present at the time of decoupling, just half a million years after the creation of the universe.

COBE's radiometers found the brightness distribution characteristic of the motion of the Milky Way galaxy with respect to the background radiation. Based on red-shift measurements of numerous galaxies, some astronomers now think that the 600-kilometer-per-second velocity of the Milky Way may arise from a gravitational tug toward a large local supercluster of galaxies called the Great Attractor. Roughly a hundred million light years away, this huge concentration of matter holds about ten thousand times as much matter as the Milky Way galaxy. Beyond the Great Attractor, there seems to be an even bigger attractor called the Shapley concentration, containing about ten times more matter still. Such large concentrations of mass suggest that the mass density of the local supercluster may be at the "critical" value, the minimum density needed for the universe to be closed. If this density were typical of the entire universe, it would have to be finite, with enough mass to make it eventually collapse due to its own gravity.

In 1992 the COBE team announced the discovery of temperature variations in their microwave map of the sky that appeared to be cosmological, not merely local. In effect, they had taken a snapshot of the universe just after the Big Bang and found what they called "the largest, most ancient structures in the universe, fifteen billion year-old fossils." The COBE map of the microwave sky (see photo insert) shows the dark horizontal band with the disk of our Milky Way galaxy. Above and below the band are blobs and mottled dark regions. If the COBE team has correctly eliminated the biasing effect of local matter, this structure shows clumping of matter in the early universe, half a million years after it began (although most of these blobs are noise and instrumental fluctuations). It is the first evidence that the early universe was not completely uniform in temperature. Many cosmologists felt relieved when the COBE team announced their discovery of temperature differences in the microwave background, since it gave structure to the early universe that could explain the evolution of the universe into galaxies and clusters of galaxies.

Until the late 1970s, many astronomers thought that clusters of galaxies were spread more or less uniformly through the Universe. As automated techniques for measuring red-shifts of thousands of galaxies came into use, however, this picture changed radically. Over vast sectors of the sky, galaxies seem to congregate in superclusters, forming lacy filaments, knots, chains, and sheets. Dark, apparently empty regions constitute much of space; many galaxies seem to gather around the edges of huge bubblelike structures about 150 million light-years across. Inside the bubbles are only a few irregular galaxies and little other visible matter. By contrast, our own region of space is densely packed, containing one galaxy about every million light-years. In 1989, Margaret Geller and John Huchra of Harvard University reported a surface containing thousands of galaxies that extends over 500 million light-years; it has become known as the Great Wall. Using the largest telescopes and most sensitive electronic detectors, astronomers can now survey galaxies billions of light-years away. They have discovered galaxies at distances of over 5 billion light-years. Whether uniformity exists on such scales is not yet clear. Some surveys seem to suggest a regular spacing of galaxies on a scale of about 400 million light-years; others see clumps, strings, and voids out to a few hundred million light-years, then no larger structure. Could COBE see the ancestors of any of these structures?

It is not apparent from the map, but the temperature differences that COBE detected are about 200,000 times smaller than the 2.7-degree temperature of the radiation (about 15 microkelvins, or 15 millionths of a degree). In order to plot the map and reveal the ancient structure hidden in their data, the COBE group first subtracted a constant intensity corresponding to 2.7 degrees Kelvin. Then they subtracted the pattern resulting from the motion of our galaxy in space—with respect to the cosmological shell of plasma that emitted the radiation. After the subtraction a speckled pattern of variations remained, hot spots and cold spots with a minimum angular size of about 10 degrees (or one thirty-sixth of the way across the COBE map, which covers all 360 degrees of the sky). No structure we see in the sky today, not even the Great

Attractor or the Great Wall, is that large in angular size. To detect the primordial predecessors to these structures, if they exist, microwave astronomers will have to improve their sensitivity beyond even the amazing level achieved by COBE, so that they can find temperature variations of about 1 angular degree in size.

Theories of galaxy formation must presuppose large amounts of dark, invisible matter in order to provide the intense gravity needed to initiate clumping. Most of the matter in the universe, astrophysicists believe, has yet to be discovered. Luminous stars and the galaxies they populate are believed to comprise but a fraction of the total. The form of the so-called *missing mass* or *dark matter* is unknown, but its gravitational effect is like that of any other form of mass, only greater. According to general relativity, the force of gravity red-shifts radiation coming from unseen clumps of this missing matter. Thus, COBE's primordial ripples may reveal residual lumps of unseen matter. The magnitude and form of these residual lumps (specifically, the number of lumps of each size) are consistent with predictions of the inflationary-universe version of Big Bang theory. COBE's ripples therefore map the distribution of matter in the early universe. Today, after billions of years of expansion, they could have become enormous regions of space with slightly increased density of galaxies.

As more COBE data accumulates, our microwave maps of the universe should improve. Features now hazy could sharpen, once the effects of obscuring microwave emissions of the Earth, Sun, and planets are accurately subtracted. Measurements at the South Pole, where interfering water vapor is very low, will give additional information on small-scale lumpiness. The development of more sensitive microwave receivers will allow new balloon-borne measurements to contribute as well. Eventually, an improved version of the COBE satellite may be lofted into space, enabling us to take sharper pictures of the ancient universe and perhaps see the ancestors of today's visible superclusters of galaxies.

To image the universe between the beginning of the Big Bang and half a million years later requires radically different techniques from those based on microwaves. Prior to half a million years, when plasma "froze" into atomic hydrogen and helium, the universe was opaque to electromagnetic radiation at all wavelengths. All the way back to the first few minutes, there was simply too much scattering of radiation by unbound electrons for any useful information to survive. This means that we cannot use light or microwaves or X-rays or even gamma rays to "see" what the universe was like before it was half a million years old.

But there may be other ways to "see" farther back. Weakly interacting, and therefore very penetrating neutrino particles must have existed in the early universe. After surviving a 10-to-15-billion-year journey, neutrinos may carry secrets of earlier phases of the Big Bang. Underground instruments in a mine in South Dakota have detected a few neutrinos from our own Sun (although the number detected is half that predicted in stellar theory), and other instruments caught a burst of neutrinos from Supernova 1987A. But we currently lack the means to detect enough of these ghostly messengers to decode any messages they may carry from the Big Bang.

Gravitational waves provide another potential means to pierce the veil of the earliest universe. General relativity requires the existence of waves in a gravitational field, much as waves of light exist in an electromagnetic field. In principle, we know how to detect them, using large blocks of metal and superconducting electronics. They should be created in supernova explosions. The violent explosion of the Big Bang, cosmologists think, would have transferred substantial energy to such waves, maybe even most of its energy. But despite strenuous efforts spanning three decades, physicists have been unable to detect any gravitational waves. If we could ever discover enough such waves to image the earliest stages of the Big Bang, we might capture a snapshot of creation itself.

CHAPTER TWENTY

Of Matter and Antimatter

During its explosive first few minutes, our universe evolved from an enigmatic state inaccessible to current physical theory to a composition similar to what we observe now. In a series of transformations brought about by rapidly decreasing temperature, the particles of ordinary matter formed, accompanied by intense electromagnetic radiation. These changes can be compared with the "freezings" and condensations of ordinary matter. The heavier chemical elements, as we saw in Chapter 12, would not be formed until much later, after stars were born. Whenever particles of matter are created, we know from laboratory experiments, particles of an electrically reversed sort called antiparticles are also created, and in exactly equal numbers. These antiparticles and any possible antimatter made from them do not seem to be part of our everyday world.

Exactly what are antiparticles? Is there really such a thing as antimatter? Although their science-fiction flavor refuses to dissipate even after one has spent hundreds of hours looking at their tracks, antiparticles are routinely produced in high-energy physics experiments. (Tracks form in particle detectors as passing charged particles ionize atoms of the detector material.) One antiparticle common in experiments is the *positron*, or positive electron. Another is the *antiproton*, the negative version of the proton. *Antineutrons* are also common, but lacking charge, they leave no tracks. In fact, virtually all the particles discovered in experiments with particle accelerators have well-known antiparticles. The symmetry between matter and antimatter in the submicroscopic world suggests that the early universe was half antimatter. This conclusion has many important consequences for cosmology. If antimatter existed today other than as isolated antiparticles, it would be composed of antiatoms, with antiprotons and antineutrons at their core, surrounded by a cloud of positrons. From a distance, such antimatter would look and behave exactly like ordinary matter.

In the presence of matter, however, antiparticles have very short lives. Their brief existence inevitably ends in total annihilation. Antimatter annihilation is the most powerful release of energy known, liberating at least a hundred times more energy than even the fusion reactions of a thermonuclear bomb. This is because *all* the energy given by Einstein's $E = mc^2$ is released, compared with the tiny fraction set free in typical nuclear reactions. Consider a 60-kilogram human shaking hands with his antimatter counterpart. The resulting explosion, equivalent to several hundred thermonuclear bombs, would replace the largest metropolis with a smoking crater.

After a particle and its antiparticle are annihilated, all that remains is radiation—gamma rays, to be precise. It was exactly the annihilation of antiparticles that gave rise to the radiation that dominated the universe after about $t = 1$ second, or 1 second after the Big Bang began.

The discovery and explanation of antiparticles was one of the greatest triumphs of modern physics. In the 1920s, the brilliant English physicist P.A.M. Dirac was searching for a mathematical description of fast-moving electrons. He realized that he needed to combine the special theory of relativity with the quantum theory of wave mechanics. Working out the details, he was able to explain many of the electron's properties, such as its spin. He noticed, however, that the resulting equations had solutions demanding positive electronlike particles as well as negative electrons—what became known as antiparticles. At first they seemed even more mysterious than our present concept of antimatter because they were erroneously thought to have negative energy.

Not even Dirac accepted the conclusions demanded by his own equations. He thought the equations were incomplete. If correctly modified, he felt, they would "predict" the proton because no positive particle with the mass of an electron existed. But he was forced to change his mind in 1932. In those days, physicists depended on cosmic rays to initiate high-energy collisions. While studying cosmic ray interactions in a cloud chamber, Carl Anderson of CalTech found some electronlike tracks that curved oppositely, as a positively charged particle would. These tracks had in fact been left by positrons. In 1937 he discovered another new particle, with a mass 207 times greater than the electron. This particle, which became known as the *muon,* also turned out to appear in positive and negative varieties, each the antiparticle of the other. In 1947, physicists Cecil F. Powell and Guiseppe Occhialini were again studying cosmic rays when they discovered a new particle with a mass 273 times the electron, the *pi meson* or *pion.* Once more they found symmetry—a positive pion and its negative antiparticle with the same mass. As soon as a sufficiently powerful particle accelerator was built at Lawrence Berkeley Laboratory in the 1950s, Emilio Segre and Owen Chamberlin were able to discover the much more massive antiproton. Soon after, colleagues found the antineutron. By the time the present authors got into experimental particle physics in the 1960s, a dozen or so new particles were known, each with its antiparticle.

In attempting to understand the strong interactions of certain of these more unusual particles, such as K mesons and Xi hyperons (known to physicists as "strange" particles), the authors spent endless hours peering at tracks. We saw antiparticle tracks every day, in images captured with a huge hydrogen bubble chamber. The high-energy beam coming into the chamber was itself composed of negative K mesons—antiparticles to the positive K mesons. Entering the chamber in a gentle arc formed by an intense magnetic field, these "strange" mesons would collide with hydrogen nuclei (protons) and often produce a heavy neutral particle called a *lambda hyperon* or Λ^0. The lambda revealed itself only indirectly, by a long V pointing back to the vertex where the K^- had seemingly vanished, or by producing a shower of charged tracks. Being unstable, the lambda particle had within a distance of a few centimeters (and in only about 10^{-10} seconds—that is, a hundred trillionths of a second) decayed into an ordinary proton and a negative pion, antiparticle to the positive pi meson.

Only a detailed computer analysis of the angles and curvatures of the tracks showed exactly what had occurred in each picture. These "fits" depended completely on relativistic calculations of the particles' energies and momenta. If special relativity had not held for such events, our data would have showed relativity to be clearly wrong. Our measurements of the lifetimes of these particles illustrated perfectly the stretching of time experienced by extremely fast-moving objects. Especially common in the photographs were a pair of graceful spirals curving in opposite directions. These revealed the formation of a positron-electron pair from an invisible gamma ray—literally, the creation of matter and antimatter from pure energy.

From many experiments involving antiparticles, two striking regularities have emerged—both with spectacular implications for the early universe. When *leptons* (low-mass particles such as electrons and muons) were created, *antileptons* such as positrons were also created. Physicists invented a quantity called the lepton number (L) to keep track of the particles; L = +1 for each lepton and L = −1 for antileptons. They expressed this apparent symmetry between matter and antimatter as the law of conservation of lepton number, meaning that the total L in an interaction does not change. In other words, the number of leptons minus the number of antileptons stays the same.

Particles such as protons and neutrons are called *baryons* (meaning heavy particles). Whenever an antibaryon, said to have baryon number B = −1, such as an antiproton, appears in a high-energy experiment, a new baryon (B = +1), such as a proton, appears too. The law of conservation of baryon number summarizes these observations of symmetry. Baryons and antibaryons are no longer considered elementary; rather, they are understood to be composed of three *quarks*, particles with electric charge one-third or two-thirds. Although quarks are essential to modern particle theory, they have never been observed to come out of the proton.

Now we can see why cosmologists believe the early universe contained so much antimatter. The very high temperatures and energies prevailing during the first second made it possible for particle-antiparticle pairs to be created from superintense radiation or from other high-energy collisions. If baryon number and lepton number were completely conserved in those interactions, the amount of antimatter would exactly match the amount of matter. But an exact match creates a troublesome contradiction with reality: Our universe seems to consist almost entirely of matter!

One way to resolve this dilemma is to suppose that somehow matter separated from antimatter and remained detached. Maybe whole antigalaxies, with antistars composed of antimatter, were formed in the later expansion of the universe. An antimatter galaxy wouldn't necessarily look any different from a matter galaxy—even the Andromeda galaxy might be antimatter. Half the universe could be antimatter—or could it?

As galaxies move through the interestellar medium, they sometimes collide, passing through one another without many stellar collisions but with much mixing of gas and dust. A collision of an antimatter galaxy with a matter galaxy would produce intense annihilation, creating detectable amounts of annihilation radiation at the interface. Astrophysicists have reported many strange phenomena in recent years, but not antimatter annihilation. Perhaps matter and antimatter somehow stay apart in the universe. A water drop remains intact on a hot frying pan for a surprisingly long time—a layer of steam acts as a barrier. In the 1960s, Nobel-prize-winning plasma physicist Hannes Alfvén suggested that something similar could happen between matter and antimatter in distant regions of space, but his idea has remained entirely speculative.

The cosmic rays bombarding Earth, consisting mostly of energetic protons (hydrogen nuclei), also contain minute amounts of virtually all the other chemical elements, from helium to uranium. As we have seen, astrophysicists think these nuclei are accelerated in supernova explosions in distant parts of the Milky Way galaxy, as well as in remote galaxies. During the 1970s, Luis Alvarez, Andy Buffington, Charles Orth, and George Smoot at Lawrence Berkeley Laboratory, as well as Bob Golden (another ex-Alvarez student) at the Johnson Space Center in Texas, searched diligently for antimatter nuclei in the cosmic rays. Since an antimatter nucleus would be negatively charged, it would curve in a magnetic field oppositely to a matter nucleus, giving it a distinctive signature. It would also annihilate spectacularly. Yet among thousands of nuclear tracks examined, not a single confirmed example of an antimatter nucleus has been found. Bob Golden eventually did find antiprotons, but these are easily explained as the result of collisions between cosmic ray nuclei and interstellar gas.

Of the many attempts to find evidence for bulk amounts of antimatter in the universe, most have shown that bulk antimatter does not exist. (The other attempts have been inconclusive.) The obvious if disappointing conclusion from this failure is that there isn't any antimatter in the universe. Apparently, the annihilation that ended shortly after $t = 10$ seconds cleared the universe of antiparticles, leaving only particles of matter. Our failure to find antimatter leaves us, however, with another, more profound dilemma: How could there be an excess of matter over antimatter if baryon number and lepton number were conserved in the creation of the universe?

Physicist Andrei Sakharov, famed as one of the chief Soviet scientific dissidents, was also the father of the Soviet H-bomb. As early as 1967, he pointed out that for the universe to evolve with an excess of baryons over antibaryons, three symmetry laws would have to be violated in "nonequilibrium" conditions, which exist when the temperature of the universe is dropping rapidly. The dropping temperature ensures that particles formed in decays can't react with each other and re-form their parents. The three symmetry laws Sakharov identified included baryon number conservation and two other laws involving charge and left-right symmetry (C and CP conservation).[1] The laws hold in *nearly* all interactions except in the decays of a particle called the K_L^o (which is its own antiparticle), when they *are* violated—the K_L^o decays more often to a positron than to an electron. In 1967 there was no theory to explain these surprising violations, so Sakharov was unable to describe a complete scenario for matter to win out over antimatter in the early universe.

Within a decade, the grand unified theories (GUTs) resolved this dilemma, bringing together the weak, electromagnetic, and strong forces. Abdus Salam and Steven Weinberg unified the weak and electromagnetic forces, while Sheldon Glashow clarified the relationship of the strong and electromagnetic forces. All three shared the Nobel prize for their work. The grand unified theories include extremely massive particles called X-bosons, which existed in the extremely high-temperature conditions of the universe before $t = 10^{-35}$ seconds. These monster particles are neither matter nor antimatter and neither baryon nor lepton. When they decay, they break the usual rules. Thus baryon and lepton number conservation can be violated at sufficiently high energies, when the distinction between strong and weak interactions dissolves.

1. C, or charge conservation, refers to the fact that electric charge is neither created nor destroyed in any known process. P, or parity conservation, which is followed only in some particle interactions, means that the interaction does not distinguish between left and right. Even though P conservation is violated in many particle decays, the product CP is not.

How were the particles and antiparticles of the universe created in the first place? Conditions in the earliest universe, say before 10^{-43} seconds, were profoundly different from those today. The overall energy density was enormous. Spacetime was exploding rapidly and may have been tightly curved, although we don't know whether it curved back on itself, as a closed universe would require. Suppose there were no particles at first, just a vacuum. According to particle theory, random fluctuations can bring particle-antiparticle pairs into existence directly from the vacuum. There's no violation of conservation of energy in this, so long as the pair annihilates before its existence can be detected. Perhaps the universe itself is such a fluctuation, an unstable accidental phenomenon that only seems long-lived to us because we don't understand time well enough. As yet we can't calculate a rate of particle creation from the vacuum. But, according to general relativity, mass and/or energy cause spacetime to be curved, while the curvature of spacetime determines the trajectories of energetic particles. Maybe the energy about to emerge from the vacuum created curved spacetime, which simultaneously guided mass/energy into existence. In other words, in a mere 10^{-43} second or less, the universe gave birth to itself.

By $t = 10^{-35}$ seconds, the universe contained a primeval soup of leptons and quarks. Particles, antiparticles, and photons were present in comparable numbers, with a slight excess of matter over antimatter. A rough equilibrium prevailed in this soup, with as many particles and antiparticles created as destroyed. Sometime between $t = 10^{-35}$ seconds and $t = 10^{-6}$ seconds, the quarks and antiquarks "condensed" or "froze" into ordinary nucleons and antinucleons. At around 10^{-4} seconds, when the temperature was about 10^{12} degrees, the energy available from average collisions was no longer enough to produce nucleon-antinucleon pairs. As annihilation continued unabated, the number of strongly interacting particles was greatly reduced until only the slight original excess of matter over antimatter remained.

By $t = 10$ seconds, there was no longer enough energy in the radiation field to create positron-electron pairs. Nearly all the existing positrons annihilated, producing two gamma rays for each annihilation and leaving only a small residue of unannihilated electrons.

Our present universe is entirely composed of these "leftovers" of matter. Today there are about two billion photons for each nucleon. These massless photons of radiation are basically a result of annihilation in the early universe, although they may have been absorbed and reemitted many times. Thus the original asymmetry between matter and antimatter must have been about one part in a billion. So not only are we made of matter cooked in stars, but the ingredients from which the stars formed were just a tiny dash of debris left over from a universe once a billion times more massive.

After t = 10 seconds, with nearly all the massive particles annihilated, most of the energy in the Universe was in the form of radiation, with comparable amounts in massless photons and in neutrinos. Since the neutrinos responded only to the weak force, they barely interacted with anything else. Only if it turns out that neutrinos have a tiny mass, as some physicists have speculated, could they have played a major role in the universe as it later evolved. If they do have mass, neutrinos may turn out to constitute a substantial fraction or even most of the mass in the present universe.

Although for half a million years after t = 10 seconds the universe was dominated by radiation, the relatively small remnant of matter had not lost its potential for excitement. Up until about t = 100 seconds, nuclei of deuterium (consisting of a proton and a neutron) and helium (two protons and two neutrons) could form in fusion reactions, but they would immediately break apart due to collisions with fast-moving surrounding particles. Over the next few minutes of cooling, the balance shifted toward stability. This transition could be described as a rapid burning, since fusion reactions on balance liberate energy. Or it could be called a condensation, since 25 percent of the known matter of the universe was then tightly locked up as helium.

Between t = 10 minutes and t = 500,000 years, the universe was an expanding plasma of electrons, hydrogen, and helium nuclei bathed in radiation (photons). The relative numbers of photons and nuclei did not change, but the total energy of the radiation plunged as each photon was red-shifted to longer and longer wavelengths. By the end of this "radiation era," the amounts of energy in radiation and matter were comparable. The temperature had dropped below 6,000 degrees. Hydrogen atoms could now form from protons and electrons without being broken apart by collisions. With the disappearance of most charged particles, the photons (which interact much more weakly with neutral atoms than with unbound electrons) became decoupled from matter, and the universe was now, for the first time, transparent. These photons would later be further red-shifted to become the microwave background radiation.

Helium produced in the few-minute-old Big Bang has remained in that form to the present day; most of it can be found in the interior of stars. The other 75 percent of known matter—except for a sprinkling of heavier elements—consists of hydrogen in stars or in the interstellar gas. The 25 percent helium and the approximately 3-degree Kelvin temperature of the microwave background radiation provide a crucial consistency check for Big Bang theory. A 3-degree temperature today extrapolates back to a temperature (in the billions of degrees) at which 25 percent of matter would be converted to helium. This cross-check, along with the observed expansion of the galaxies and the near-uniformity of the microwave radiation, constitute the bedrock upon which rests our reconstruction of the early universe.

CHAPTER TWENTY-ONE

Universes Finite and Infinite

We have traced the explosion of our universe from the Big Bang to the present and shown how those events led to the possibility of human life. But will the ongoing expansion simply continue? Will humans endure and keep on evolving? Is there an endpoint for the universe, or will it go on forever? These questions are closely tied to another issue we have skirted somewhat gingerly: Is the universe finite or infinite? And they relate to the curvature of space: If, as general relativity stipulates, space is curved, *how* is it curved? Is its large-scale geometry like what we learned in high school, or is it profoundly different?

Although nearly all cosmologists today work within some version of the Big Bang theory, there is no consensus on the answers to these questions. On one point, however, there is agreement: The expansion must be slowing down. All matter is attracted by the force of gravity to all other matter, which inevitably leads to a deceleration of the expansion. A simple analogy can clarify this point. If you throw a ball straight up in the air, gravity slows it down while it is on its way up, then stops it completely for a single instant, then brings it hurtling back to Earth. Perhaps the expansion of the universe will eventually slow all the way to zero and then reverse, as gravity finally begins pulling the galaxies inward. But could this really happen? Is it possible for the universe to collapse?

Lingering another moment on our analogy, you could conceivably throw a ball upward at a speed of over 11 kilometers per second. In that case, the ball would completely escape the Earth's gravitational pull. It would have exceeded the "escape velocity" for the Earth. For the universe, the key to its fate hinges on a single number, its mass density. If its mass density is high enough, then gravitational attraction will pull the universe inward (that is, force it to collapse), just as a ball thrown upward at less than escape velocity must return to Earth. If the density of the universe is less than some critical value, however, it will expand forever. In an in-between case, the receding galaxies would eventually reach a relative velocity of zero when they are infinitely far apart. Our modern theory of gravity, general relativity, can accommodate each of these situations. One currently favored variant, the inflationary universe, requires that the density be almost *exactly* the critical value.

In observational astronomy, the density of the universe is still an unanswered question. The amount of mass in visible objects like stars and galaxies doesn't seem nearly enough to close the universe—to stop it from expanding forever. There isn't even enough mass to explain the motion of clusters of galaxies; although they act as if they have enough mass for gravity to hold them together, the mass density seen in visible stars suggests otherwise. This puzzle has triggered a determined search for so-called dark matter, in the form of either invisible objects like brown dwarf stars and black holes or unknown elementary particles. It is very difficult to measure the mass of weakly interacting particles like neutrinos, but if they had a rest mass of about 10^{-7} proton masses, then neutrinos given off early in the Big Bang would account for enough dark matter to eventually reverse the expansion. Measurements to date, however, indicate that the neutrino does not have this much mass.

Simple models within the general-relativistic theory of gravity make different assumptions about mass density. They even demand different curvatures of space, which means they have different large-scale geometries with curious properties.

According to one of these models, known as the closed model, the universe will eventually collapse. The closed model, which originated in 1922 when the Russian mathematician Alexander Friedmann discovered several solutions to the equations of general relativity, assumes a finite number of galaxies in a finite volume of space. This seems logical enough, but watch out—there are also no boundaries to this volume, nor does it have a center. In every direction space appears the same. Furthermore, we have no simple way to visualize this space, which curves in three spatial dimensions. We can't conceive this space, any more than two-dimensional creatures, like our water-wave creatures from Chapter 16, could visualize a third dimension perpendicular to their ocean. Even if their world were the surface of a large sphere, which can be described by the two dimensions of latitude and longitude, they would probably think they lived on a flat planet.

Questions about the properties of space have interested scientists, mathematicians and writers for a long time. More than a hundred years ago, Edwin Abbot first introduced the concept of two-dimensional creatures struggling to understand a three-dimensional world in his science-fiction classic *Flatland.* (Readers interested in a more detailed but not highly mathematical treatment will also enjoy George Gamow's classic *One Two Three . . . Infinity.*)

Suppose the water-wave creatures could observe each other by means of light that traveled in the shortest possible path on the surface of their planet, an arc of a great circle. Their world is finite. They've been brought up to think that their world is *flat;* they can't understand how their world could look the same in all directions unless they were at the center. But no one of them is at the center of anything. Their world as *we* see it from the outside is curved in a third dimension unobservable to them. Similarly, if Friedmann's closed universe is the right model, *our* world is curved in a fourth dimension unobservable to us. This fourth spatial direction is useful mathematically in discussing the model, but hardly anyone can visualize it, and no one can tell whether it really exists or not. A few mathematicians, like Bill Thurston, say they *can* visualize the fourth dimension. Considering the large number of important theorems he has discovered, his claim is probably correct.

How can creatures on a curved surface discover that it is not flat? One way is to draw circles of larger and larger size. On a plane, the circumference of a circle divided by its radius is 2π = about 6.28. This relationship would hold on the creatures' planet as long as the circles were small, but for circles that covered a substantial chunk of the planet's surface, the circumference-to-radius ratio would decrease dramatically. Imagine one creature remaining at the north pole of his world while a companion, heading due south, pulls out a long measuring tape. By the time the companion gets to the equator, he's pulled out enough tape to go one-quarter of the way around the planet. Now suppose he walks all the way around the planet, staying on the equator. He's drawn a circle with a radius one-fourth of its circumference—the ratio of circumference to radius is 4, rather than 6.28. If the companion creature pulls out enough tape to get to the south pole, then tries to walk in a circle at the same latitude, he could even measure a ratio that approaches zero—since the circumference is close to zero but the "radius" is 12,000 miles! (A tape stretching from the North Pole to the South Pole is about 12,000 miles long, half the circumference of Earth.)

On a plane, in the familiar geometry of Euclid, the sum of the three angles of a triangle is 180 degrees. But on the surface of a sphere, large triangles have sums of angles much bigger than 180 degrees—even 270 degrees or larger. To understand this, take a globe of the Earth or a basketball. Draw a triangle from the North Pole to the Equator, go one-quarter way around the globe, then go back to the pole. The triangle has three 90-degree angles! (You can find triangles with bigger angles—the limit is 540 degrees for the whole triangle.) If the water-wave creatures did this and were good enough at geometry, they could calculate, from their measurements, how curved their world is.

What if the creatures' finite world is expanding, as ours seems to be? The creatures' planetary surface is carried outward as time passes, but this is difficult for the creatures to understand because the change in position of each of them (let's assume they are standing still) is in the third, unobservable dimension. They think expansion means motion on their familiar surface, just as we are tempted to think the expansion of our universe means that galaxies have real velocities away from us. Actually, in a relativistic model, the separation of galaxies is due solely to the expansion of the space between them, not to any velocity that they "have." (Strictly speaking, the red-shifts of the galaxies are also due solely to the expansion of space and not to their velocities.)

In the closed relativistic model, the three-dimensional universe is carried "outward" to new positions in an unobservable fourth dimension. Its volume increases smoothly in all directions. It is perfectly feasible for the total volume of space to increase in this context because with gravity varying inside it, there's no reason for volume to be constant. The curvature of spacetime produces expansion, while the distribution of matter and energy makes spacetime curve a certain way.

All this means that the fourth dimension has a certain *radius of curvature,* comparable to the radius of the spherical planet of the water-wave creatures or to the radius of the Earth. As we saw in Chapter 16, this radius is the distance between any point in our three-dimensional space and the "center" of the fourth-dimensional space. For the universe, this radius of curvature can be referred to as the *radius of our universe.* Furthermore, the radius is increasing with time. If you traveled straight in any direction to a distance equal to 2π times this radius—at infinite speed—you would come back to where you started. So the farthest points away from you, the "opposite ends" of the universe, in all directions, would be π times this radius. Unfortunately, this feat could not actually be accomplished in Friedmann's model of a closed expanding universe, because the length of the trip is always more than the speed of light times the age of the universe.

Furthermore, in his model the galaxies aren't moving spatially (relative to each other) at all—they are carried outward on a specially chosen coordinate system that moves with them. But they are moving forward in time—that is, they are getting older. The universe expands not because the galaxies are flying apart, but because the radius of curvature of the universe is increasing. We say that space is expanding and is increasing the separations between galaxies. Their red-shifts are due to space expansion, not to actual recession speeds. Galaxies that were once on the far side of the universe from us will always be on the far side, and we will never be able to see them—not as long as the expansion continues. We can't see to the "opposite end" of the closed universe because light doesn't travel fast enough to make this possible. And we certainly can't think of those farthest-away galaxies as having sped away from us, because they would have had to travel faster than the speed of light to get where they are now.

What, in this picture, is the fate of the universe? Expansion would slow gradually for billions of years. Then in some future era, say 50 or 100 billion years from now, it would reverse entirely. The drama of the Big Bang would play out backward as the Big Crunch. There would still be galaxies and bright stars, since new galaxies keep forming from the intergalactic gas due to gravitational attraction. But there would be far more burnt-out, dead stars than there are now. If astronomers were left alive anywhere (there would likely be none on Earth because life on our planet would have been destroyed when the Sun became a red giant star), they would observe blue-shifts resulting from the contracting space between then-existing galaxies. And they would finally be able to see those farthest galaxies, since enough time would then have elapsed for light from the galaxies to reach them.

As the universe contracted, its gravitational potential energy would be converted to kinetic energy and eventually, by myriad collisions, to heat. All matter would be compressed. The microwave background radiation, having previously cooled to less than one degree Kelvin, would eventually reach 6,000 degrees again. The universe would now have less than a million years left. Since its total energy would have been conserved in the collapse, the various "freezings" and condensations of nuclear matter would be reversed, now in cataclysmic periods of melting and vaporization. First the universe would reenter its opaque, radiation-dominated plasma stage. A few minutes from the ultimate Crunch, the universe would again be an incredibly hot nuclear soup. All nuclei more complex than a single proton would be broken apart.

Finally, something extremely dramatic would happen. Ten seconds from the final instant, the energy density of the universe would again be large enough for electron-positron pairs to be created everywhere. At $t = 10^{-4}$ seconds from Crunch time, nucleon-antinucleon pairs would again appear in dominant numbers. A little later, quarks would replace the baryons and mesons. By 10^{-35} seconds, the primeval quark-lepton soup would be re-created.

We do not know whether the original matter-antimatter asymmetry would also be re-created. Nor do we know what would happen next. Would we return to the brief era of the X-boson, postulated by grand unified theory? Would the universe then simply disappear?

One popular speculation about the closed universe has been that it would bounce and explode again in a new Big Bang. In that case the fate of the universe would be cyclic, with no beginning or end. But there is no compelling physics to indicate that an imploding universe would bounce, and simple versions of the bounce violate the equations of general relativity. Most speculations about the cyclic universe avoid considerations of physics altogether and emphasize its similarities with Hindu and other ancient cosmologies. In his book *The Big Bang*, University of California astronomer Joseph Silk discusses some intriguing constraints on cyclic models. Each expansion and subsequent collapse would produce radiation as starlight, radio waves, and X-rays. During the collapse this radiation would all end up as black body radiation. Unless the collapse proceeded to the point where most of the radiant energy was transformed into particle-antiparticle pairs, the radiation would accumulate. Since we only observe a certain amount of radiation now, this sets a limit on the number of bounces that could have occurred, about a hundred bounces. Such a limit, if it exists, considerably reduces the charm of the closed, rebounding universe model. Exactly what mechanism would produce a bounce either at the end of the radiation era or later, we don't know.

What if the Big Crunch continued right down to the singularity, the point where the energy density of the universe becomes infinite? General relativity cannot be relied on to tell us what would happen then, since its equations "blow up"—become infinite. But the implied tiny time interval suggests that quantum phenomena would become important. General relativity, however, is not a quantum theory; nor is there any other successful quantum theory of gravity. Indulging in wild speculation, we may suppose that at the point of total collapse, *space and time would simply end.* The universe would then have gone full circle—from nothing to nothing.

One nagging problem with the closed universe model is its difficulty in accounting for the amount of deuterium now observed. As we mentioned earlier, some deuterium was made during the thermonuclear detonation that occurred at about $t = 100$ seconds. However, a closed universe model requires such a high matter density at that time that deuterium would simply burn (to form helium), and none of it would survive. Although there are some complicated ways to escape this problem in a closed universe, they lessen its appeal.

Amusingly, the closed universe technically is a black hole. Its mass has distorted spacetime enough that it has folded back on itself, ruling out the possibility of light or anything else escaping. Of course, it's meaningless to talk about an *outside* of the universe; as Gertrude Stein once said of Oakland, California, "There's no there there."

What about models of an open universe? An open universe would be unbounded, infinite, and expanding forever. The geometry of space for the open model, also discovered by Friedmann, is once again beyond most people's ability to visualize. If the two-dimensional analog of the closed model is the surface of a sphere, the two-dimensional version of the open model is a saddle shape. Circles on the saddle have a circumference-to-radius ratio *greater* than 2π rather than less, as on the surface of a sphere. Even stranger, the sum of the angles of a triangle on the saddle surface is less than 180 degrees. As mathematicians say, the curvature of space is positive for the closed model and negative for the open model.

It's awkward to rely on a surface that has edges to visualize one that, like the open universe, does not. You may imagine that the saddle extends to infinity, but bear in mind that the saddle is just an analogy—no one is suggesting that an open universe looks like a saddle, any more than the closed universe looks like a sphere. Despite the unusual saddle shape, the geometry of spacetime in an open universe is less bizarre than in a closed one. The mass density of the open universe is much smaller, so spacetime is not as sharply curved as in the closed universe. But the number of galaxies in the open universe, and the total mass, are infinite. As the open universe expands, the mass density drops until the effects of space curvature become negligible. Physics then follows the rules of Einstein's special relativity rather than general relativity.

The future of an open universe is cold. Since the expansion has no end, all stars will eventually run out of nuclear fuel and die. Although some will die explosively and cast their matter out into interstellar space, the density of the gas and dust will drop to the point where new stars and new galaxies can no longer form from gravitational attraction. In the absence of young stars, the universe will be overtaken by darkness. Matter will cool to absolute zero. Huge black holes will form from the collapse of galaxies. (Many astrophysicists believe that there are already giant black holes at the cores of galaxies.) If the proton is unstable, as required by grand unified particle theory, with a lifetime of about 10^{32} years, all matter will disintegrate. But will the open universe eventually reach a condition where nothing whatsoever happens except endless expansion? Perhaps not. Freeman Dyson has shown that life could evolve more rapidly than the expansion, so that as the universe cools, the evolved creatures would develop new knowledge (including how to do without sunlight) and continue to find life *interesting.* So even without sunlight, our future may be bright!

Finally, in the third model, the universe may be right on the border between closed and open. In this model the large-scale, cosmological curvature of space is zero.[1] The geometry of space is Euclidean just as is taught in school: The circumference-to-radius ratio of a circle is 2π, the sum of angles in a triangle is 180 degrees, and so on. To visualize the model, we return to the two-dimensional analogy, but this time it is just a flat, unbounded plane. The ultimate fate of a flat universe would be much like that of an open one: a cold sea of radiation at essentially absolute zero that barely stops expanding after an infinite time.

Apart from our aesthetic preferences, which we allow to color scientific judgment only at our peril, what tools do we have to decide which of these models is accurate? One potentially powerful tool is the plot of Hubble's law called a Hubble diagram. Hubble plotted the red-shift (equivalent to the recession velocity) of galaxies as a function of their distance. If the universe is expanding at a constant rate, galaxies lie on a straight line on this plot. But in none of the models does the universe expand at a constant rate, due to gravitational deceleration. Rather, each of the various models predicts a slightly different relationship. All the models predict a nearly linear relationship at small red-shift, where most of the data are; but the predictions diverge at high red-shift, where there are knotty complications. The upshot is that despite decades of effort, the galaxy red-shift data are ambiguous, and all three of the models are quite possible.

The actual mass density of the universe cannot be as large as, say ten times the critical density needed to close the universe; if it were, the expansion would be slowing down much faster than the observed rate. The luminous matter in stars and galaxies, plus the gas we can detect gravitationally, comes to only about one-tenth the critical density. So we know the actual density to within a factor of 100 (the range from 1/10 to 10).

The inflationary hypothesis of the grand unified theories of matter leads to a very definite preference for the closed-open boundary model. After the period of inflation, in which a very rapid expansion ends with an enormous release of energy as particles with mass, the more-or-less-well-understood nuclear interactions dominate the universe. Calculations show that the mass released at the end of the inflation period is

1. Space is still curved in the local region near mass. Even if the cosmological curvature is zero, the curvature of spacetime near the Sun still is responsible for the circular orbit of the Earth.

exactly the mass needed to close the universe! Therefore the inflationary model requires that the universe hover at the boundary between just barely closed and just barely open. For that mass density, spacetime is flat.

Inflation seems to us to be a necessary part of the Big Bang picture, but it would be dangerous to accept the cosmological model it favors on purely theoretical grounds. Exciting new developments in the field of supernova research hold promise of breaking the observational logjam and distinguishing between the three models. All the Big Bang models have much in common; in each of them the universe is everywhere and always was. In each there was no special place of creation. Creation happened everywhere. Looking outward, we look backward in time at a small fraction of the entire universe. As the universe ages, we can see more and more of it, as galaxies come over the horizon at the speed of light. (The horizon is the distance light has traveled since the universe began.) If the universe is infinite, we will never be able to see more than a tiny portion of it. In any case, we will never be able to see the moment of creation itself, when

> *Time began and the Universe exploded, erupted*
> *from nothing, filled with fire and light*
> *everywhere, furiously hot and bright.*

CHAPTER TWENTY-TWO

Cosmological Candles

Any theory that claims that the universe had a beginning must be able to attribute an age to it, consistent with all astronomical data. Certainly the time that has elapsed since the Big Bang cannot be less than the known age of any object in the universe (except perhaps photons from previous bounces of a cyclic universe). Although measuring the age accurately is immensely difficult, cosmologists have reached a surprising consensus that the universe is between 8 and 17 billion years old. But as they have since the time of Edwin Hubble, they are still arguing about the exact age.

Methods for determining the age of the universe depend critically on measuring distances of remote galaxies. Scientists measure these distances indirectly by comparing the brightness of the galaxies with the brightness of objects whose intrinsic luminosity they think they know, objects that astronomers often call *candles.* In addition to distance, the rate at which galaxies are receding from us is crucial for determining the universe's age. According to Hubble's law, as we have seen, the speed at which a galaxy is receding from us is proportional to its distance from us. Dividing the speed of recession of galaxies by their distance from us gives us the rate of expansion of the universe. Thus, the faster the galaxies at a given distance are receding, the higher the expansion rate of the universe. A high expansion rate implies that the universe is relatively young, because less time would have been required for distant galaxies to attain their current large distances from us. On the other hand, a low expansion rate means that the universe is older.

Technically, the expansion rate of the universe is known as *Hubble's constant.* It is called a constant because the same value holds for all regions of space; that is, it is constant with respect to location. Astronomers variously put the constant at 50 to 100 kilometers per second per megaparsec. (A megaparsec is the distance light travels in 3.26 million years.) Since the expansion rate of the universe is slowing down, Hubble's constant decreases as time goes on. Calculating the age of the universe thus depends on the choice of cosmological model as well. The 12-billion-year age that we have been using is based on the work of Hubble's own protégé Alan Sandage, who uses the widely accepted inflationary Big Bang model and, along with his co-workers, has been arguing for a relatively low expansion rate—50 kilometers per second per megaparsec—for years. But well-respected astronomers have challenged Sandage with evidence for expansion rates almost twice as high. Rates that high tend to upset cosmologists because they imply that the universe is younger than certain stars. (The ages of these stars, however, are themselves uncertain since their determination is based on complex stellar models that cannot seem to predict the correct number of neutrinos emitted by the Sun.) To complicate the situation, astronomers have come up with a variety of other methods to determine the age of the universe—ingenious if exotic—and these yield values spanning the accepted range.

In order to pin down Hubble's constant and therefore the age of the universe, astronomers often measure the red-shift of spectral lines in a distant galaxy, which is straightforward, and the distance to the galaxy, which is a far more daunting task. Such distances can't be measured directly. To find them, astronomers observe Cepheid variable stars, the same "standard candles" Hubble used to deduce his law in the first place. According to the period-luminosity relation for Cepheid stars, the star's period of oscillation tells its luminosity. Its apparent brightness then determines its distance. Methods of measuring Cepheids took a giant leap forward in 1993, after space shuttle astronauts fitted corrective optics to the Hubble space telescope. Using the Hubble telescope, Wendy Freedman measured precise light curves of twenty Cepheid stars in M100, a prominent spiral galaxy in the Virgo cluster. She and her collaborators at the Carnegie Laboratories in Pasadena, California, found that the distance to M100 is 17 megaparsecs (about 50 million light-years).

When Freedman and her colleagues calculated Hubble's constant, its value seemed to demand (using the inflationary model) that the age of the universe be only 8 billion years. This conflicts with the 13-to-17-billion year ages generally accepted for old stars in globular clusters within our galaxy. (Some theorists say the globular cluster stars could be as young as 11 billion years, another troubling discrepancy.) Either Freedman made a mistake, or the globular cluster ages are way off, or there's something wrong with inflationary Big Bang cosmology. Could the Big Bang itself be in jeopardy?

Using adaptive optics to eliminate the effects of atmospheric turbulence, Michael Pierce and his colleagues at the University of Indiana measured Cepheids in another Virgo cluster galaxy with a precision rivaling that of the Hubble telescope. Pierce's result agreed with Freedman's. A telescope equipped with adaptive optics uses electronics to measure the effect of turbulence and then corrects it in real time by rapidly moving elements of the optical system. (Richard Muller worked on this principle in the 1970s.)

Of the many other ways to measure Hubble's constant astronomers have invented, some depend on Cepheid variable stars for calibration and some do not. One class of methods to determine the distance of galaxies uses supernovae as bright objects of presumably known luminosity. Since Type II supernovae shatter massive stars with greatly differing masses, they blaze with a wide range of intrinsic luminosities. So they do not make good "standard candles" for measuring distance. But Brian Schmidt, Robert Kirshner, and Ronald Eastman of Harvard University found a way to deduce the intrinsic luminosity of a Type II event from its light spectrum. Based on eighteen supernovae, they came up with an estimate of the Hubble constant only a little less than Freedman's.

Sandage continues to insist on a lower value for Hubble's constant, and he has made some recent measurements that back him up. He calibrated a couple of venerable Type I supernovae, recorded back in 1937 and 1972, against Cepheids in their galaxies. Type I make better "standard candles" than Type II, although there's some question just how standard candle Type I supernova events are and whether their brightness can be reliably corrected. Made with the Hubble space telescope before the astronauts installed corrective optics, Sandage's measurements required great finesse. For their part, Harvard astronomers Kirshner,

Adam Riess, and William Press came up with a way to correct the intrinsic luminosity of thirteen recorded supernovae from their light curves. Their answer for Hubble's constant and the age of the universe lies midway between Sandage's and Freedman's.

Whose answers should we accept? All these measurements have about the same overall uncertainty, some 20 percent up or down. Space telescope astronomers are planning to measure Cepheids in twenty other galaxies, while supernova hunters will eventually find many more Type I's and II's. For calibration, the supernova observers need to have some supernovae in galaxies with Cepheids as "measuring posts," but most supernovae are much farther away than Cepheids can be measured. These more distant and faster receding Cepheids have an advantage over indicators in closer galaxies: They are less likely to be significantly affected by local concentrations of mass, such as the Great Attractor, which conceivably can increase or decrease measurements of the expansion rate (although Wendy Freedman and her collaborators claim to have corrected for this effect on their data).

Shortly after Einstein introduced general relativity, he suggested a class of solutions to his equations that included a mysterious term called the *cosmological constant.* Like inflation, which forces the universe to expand more rapidly than gravity allows, the cosmological constant amounts to a kind of "antigravity." After Alexander Friedmann came up with his simpler and more elegant solutions in the 1920s, Einstein regretted his introduction of the cosmological constant and called it "greatest mistake of my career." Cosmologists now are tempted to resurrect the cosmological constant, however, because used as an adjustable parameter in a Big Bang model (with or without inflation), it can accommodate a wide range of ages for the universe. In other words, the constant can make an expansion rate as big as Wendy Freedman's consistent with an age of the universe safely in excess of the age of star clusters.

As we've seen, the inflationary model demands that spacetime be flat and that the density of the universe be just above the minimum necessary to close it. This key prediction may soon be testable using Type I supernovae as standard candles. No matter what the cause of the original explosion, and no matter what the details of its first moments, the universe's expansion must slow down due to gravity. This slowdown is called the *deceleration* of the universe. It is closely related to the mass density of the universe; the greater the density, the stronger the inward tug of gravity on the universe's mass and the greater the deceleration. A large cosmological constant could, however, negate this deceleration and replace it with even faster expansion as the universe ages. What is not yet known is just how much deceleration is taking place. If the deceleration is great, our universe is closed and will eventually collapse. If it is not so great, the expansion might continue forever. And if inflation holds, the expansion will just stop as time reaches infinity, like a rock thrown upward exactly at escape velocity.

Determining the deceleration of the universe, and therefore its mass density, is one of the great challenges of cosmology. But to do so, some type of standard candle must be observed at distances far greater than mere millions of light-years. To determine the deceleration, observations must be carried out across much of the visible universe spanning billions of light-years.

Possible standard candles for this purpose include the few types of objects that can be observed at vast distances, namely galaxies, clusters of galaxies, quasars, and supernovae. Cepheid variable stars are far too dim. Galaxies are plentiful and can be seen up to billions of light-years away, but they don't make good standard candles since they vary in size. So astronomers sometimes use only the brightest galaxy or third-brightest galaxy in a cluster as a potential candle. Even these measurements are highly uncertain, however, because distant galaxies are also very old (and therefore might not have the same intrinsic brightness as nearby, younger galaxies; moreover, galaxies sometimes merge to form much brighter ones. Based on their spectra, astronomers have concluded that distant galaxies are made of different kinds of stars from close galaxies, the nearby stars containing less matter composed of the heavy elements.

When Richard Muller began the automated supernova search, its ultimate purpose was to find supernovae to use to determine the deceleration. Now, more than fifteen years later, the project has not yet done so—but it is getting close. Saul Perlmutter and Carl Pennypacker, who have taken over the project, are convinced they can find enough such supernovae for use as standard candles to make a frontal assault on the deceleration problem. These supernovae, Type Ia, may comprise a class of standard events for several reasons. They all result from white dwarfs that have picked up matter from companion stars. All Type Ia supernovae have the same mass, about 1.4 times that of the Sun. Their light curves reveal an impressive uniformity: Observed at maximum brightness, their light output varies by less than 25 percent—fairly constant, compared with the uncertainty in observing galaxies. Just how standard Type Ia supernovae are as a group remains controversial, but their appeal is great. Discovery of only a few dozen of them at sufficiently great distances might reward the Berkeley scientists with the logjam-breaking measurement of the deceleration of the expanding universe that astronomers have been seeking for decades. But these supernovae are not easy to find. Since they are so dim, they require large telescopes to detect and track them. It is tough to get enough time at the big observatories for such a search. When an astronomer does get a block of time, the vagaries of weather can spoil the run, meaning no data or unusable data.

Whether remote galaxies or the supernovae that occur within them are used as standard candles, measuring deceleration depends on the fact that the spectra of very distant objects are shifted far into the red. Each recognizable spectral line is observed at much longer wavelength than would be observed for the same line in a laboratory experiment. Either the supernova spectrum itself or that of the host galaxy can be used to measure this shift. In either case, the bigger the red-shift, the faster the distance is increasing between Earth and the supernova.

To find the secret of deceleration, Perlmutter and Pennypacker are looking for a departure from the simple Hubble law. A supernova's curve of apparent brightness versus red-shift will depend on the deceleration of the universe, and this dependence will be greatest for the supernovae that are farthest away (and which therefore have the largest red-shifts). To calculate the theoretical curves, astrophysicists must pick a simple model of the universe and then solve the equations of general relativity. In comparing curves with red-shifts, a few very distant supernovae are potentially much more valuable than the larger number that are closer to us. Even a single large red-shift supernova several billion light years away, with a very precisely measured brightness, could potentially distinguish between models of the universe.

Supernovae at these distances are so dim they would have been undetectable using older photographic techniques. The most distant exploded at the same time the Earth was being formed—about 5 billion years ago. Even CCD cameras on the largest telescope may receive only a few hundred photons from such exotic outbursts—barely enough to count above the shadowy background of their host galaxies. To succeed, Pennypacker and Perlmutter must orchestrate a campaign with military precision. Not only does their quest demand precious nights on some of world's largest telescopes, they need political skill to organize a network of astronomers willing to cooperate on follow-up observations. Rich clusters can provide hundreds of promising galaxies per image, but discovering even one supernova per block of observing time requires imaging tens of thousands of galaxies. To analyze their images quickly, they must develop software capable of scanning hundreds of galaxies in seconds—and eliminating a variety of brightness variations that mimic real supernovae. Still, they must be prepared for long nights of scrutinizing candidates by eyeball. As of early 1995, they had discovered seven high red-shift supernovae, while surveying hundreds of thousands of galaxies. But this is only a beginning. Dozens more supernovae at cosmological distances are scheduled for discovery. We may soon solve one of the deepest mysteries of the universe—and learn its secret of deceleration.

CHAPTER TWENTY-THREE

Three Big Bangs Revisited

We started this book by asking where we came from. In its own quest to answer this question, twentieth-century science has reached many a surprising conclusion and put forth many a strange theory. Certainly, citizens of earlier centuries would find utterly astounding the accounts of megaviolence that fill the preceding pages. Still, life was somewhat less secure in earlier days; perhaps the thought that a huge invading missile from space could devastate the planet would have been less shocking then. Today's electronic media keep us continuously informed about disasters everywhere on Earth, from horrible earthquakes to fires, floods, and wars. Should an asteroid strike in the megaton range occur—like the Tunguskan event of 1908, but in a densely populated area—it is not too difficult to anticipate the media reaction. There would be saturation TV coverage, squadrons of reporters on the scene, endless interviews with scientists speculating on what still more powerful impacts would be like.

But even after absorbing the first ten chapters of this book, the reader may resist the implications of a big bang on the scale of the one 65 million years ago—a holocaust that set continents afire, plunged the world into darkness, and launched mile-high waves careening across a poisoned ocean. It is possible that after such a newsworthy event, there would be no TV crews or newspapers to cover it—not even e-mail. It is conceivable that an impacting long-period comet on the high end of the size range would totally exterminate our species. Perhaps cockroaches, ants, and hardy mammals like rats would inherit the Earth. Within half a billion years or so, intelligent life might eventually evolve again and rediscover the nature of the shooting gallery they inhabit.

A century ago, scientists had intriguing clues about the submicroscopic world of atoms and elementary particles but little real knowledge of it. Any notions about the origin of matter lay in the realm of pure speculation. Today, we have a robust understanding of nuclear physics, including detailed data about some 110 chemical elements and their thousands of isotopes. Our grasp of the ultimate nature of matter is incomplete, but physicists agree on a "standard model" of elementary particles and the forces that affect them. The model provides a strong framework to explain how subatomic particles can combine into simple atoms like hydrogen and helium and how, given the right physical conditions of pressure and temperature, nuclear reactions can weld these atomic building blocks into all the forms of matter known.

By the end of the nineteenth century, astronomers had discovered and categorized countless stars of varying colors and types, and hundreds of intriguing fuzzy patches in space. Some of the stars seemed to suffer periodic explosive outbursts called novae. Still more powerful starlike outbursts had been known since ancient times. Not only were these outbursts inexplicable, scientists had no good idea what made stars, or for that matter our Sun, shine at all. Today we know. Nuclear fusion reactions power the Sun and the stars and make it possible to glue simple atoms together to make complicated ones. Ordinary stars, burning steadily for billions of years, can synthesize only a few of the elements, not nearly enough to account for life. The powerful, mysterious, and rare "super-nova" explosions turned out to hold more than one secret of biological life. For stars themselves, they embody both death and rebirth. Under the incredibly hot, violent, pressurized conditions of a shattering massive star, exotic fusion reactions cooked the heavy elements necessary to make our world. Blasted out into interstellar space as dust and gas, this star-stuff joined other tenuous matter. Eventually, pulled inward by the irresistible tug of gravity, the recycled matter coagulated to form new stars, triggering yet another bout of element formation. The process continues to this day.

Left behind by supernova explosions, whirling ultradense neutron stars have illuminated the wispy remnants for thousands of years. Each neutron star's magnetic dynamo launches charged particles on million-year journeys across space, creating a long-lasting radiation hazard to life. These tiny but high-energy cosmic messengers intercept life wherever it has formed, and by damaging DNA molecules (along with sources of natural radioactivity on Earth and chemical mutagens), they help promote continued evolution through the phenomenon of genetic mutation. Where life prospers on the exposed surfaces of temperate planets, it remains exquisitely vulnerable to inevitable collisions with chunks of rock and ice—comets and asteroids. These impacts give evolution a mighty kick by wiping out much of what has been formed previously. Once the ruinous effects of a collision have faded away, the survivors, if any, race to fill in the niches of the disappeared. Whether such big bangs constitute a random shuffling of the deck or a step toward evolution of "higher" life-forms is unclear. On Earth, the one evolutionary record currently available makes the presently dominant inhabitants—human beings—seem superior to older life-forms, but this may be mere prejudice on our part. On the other hand, ammonites, trilobites, dinosaurs, and saber-tooth tigers created no civilizations (as far as we know), so maybe humans do have some basis for conceit.

Omitted from our account of origins thus far is the biochemical (or possibly divine) step by which mere atoms and molecules became living beings. Chemists have demonstrated repeatedly the physical mechanisms that build up some of the more complex molecules of life from simple ones. Biologists know how relatively simple molecular structures like viruses reproduce, and they have even discovered simpler ones (prions) that can cause infection without DNA. They can take living viruses apart and put them back together again. Astronomers have detected an astonishing variety of organic molecules in deep space and on comets. One meteorite—a so-called carbonaceous chondrite—contains sixteen different amino acids. All of this suggests that given enough time and enough variety in external physical conditions, simple self-replicating systems (that is, life) can form from nonlife. This could have happened here on Earth, or first life might have arrived on a comet fragment or meteorite. So far, however, no one has been able to take chemicals from the shelf, combine them somehow, and make an infective virus, prion, or bacterium. But our increasing mastery of biotechnology may make possible even this feat.

One recent discovery suggests that major surprises about the nature of life on Earth may still await us. It seems that biologists had vastly underestimated the importance of bacteria living in underground rock. Even hundreds of meters beneath the surface, in a dark and seemingly hostile environment, life prospers. There anaerobic bacteria, which do not require oxygen, live by digesting rock itself. They produce waste products such as methane, the main constituent in the natural gas we use to heat our homes and cook. So plentiful are these bacteria that they may constitute more than half the biomass on Earth—more even than the forests and jungles or oceanic plankton. Their habitat lacks many of the "amenities" that humans consider necessary for the good life, but it has one advantage over ours: In an asteroid strike, except for a major direct hit on their particular chunk of Earth, these bacteria would be invulnerable. In a sense they're immortal.

Among the fuzzy nebulae that eighteenth- and nineteenth-century astronomers discovered were some that were neither supernova remnants nor giant clouds where new stars are born. The majority, in fact, turned out to be vast aggregations of billions of distant stars—galaxies. The outward motion of the more distant galaxies held the key to the deepest mysteries of our origins, among them the puzzle of how *anything* came to be. Flying apart at speeds of thousands of kilometers per second, the galaxies constitute a huge gravitationally bound system that was once much smaller and hotter. Abundant data make it difficult to escape the conclusion that the galactic expansion we see now is the result of a gigantic explosion or Big Bang—from a point of infinitesmal size and essentially infinite temperature. Yet our best theory of gravitation, general relativity, forces us to conclude that our everyday concept of size cannot survive extrapolation back to the hypothetical beginning. Notions of a pointlike universe within some larger, presumably emptier whole are incorrect because space acquires its properties precisely from the huge mass contained within the universe. Space and time exist only within the universe. General relativity does not impose a particular model or picture of the universe on us. We do not know whether the present universe is finite in extent with a finite number of galaxies (and of finite mass) or whether it is truly infinite.

Of the various models, or solutions to the equations of general relativity, that were proposed soon after Einstein announced his theory in 1915, none can correctly describe the real universe without elaboration. We have evidence that the universe homogenized itself, or smoothed out any irregularities in density and temperature, in less time than it would have taken light (the fastest possible smoothing influence) to cross from one side to the other. The evidence is the extreme uniformity of the microwave cosmic background radiation, which left the early universe less than half a million years from the Big Bang's onset on an uninterrupted journey toward us. At that time, the universe was already more than 50 million light-years across. There is no contradiction here, because the universe can (and at first, must) expand faster than the speed of light, even if nothing material or energy-containing can move across it faster than this speed. The Big Bang expansion is simply a stretching of space, and nothing in relativity prevents space from stretching faster than the speed of light.

There are several ways out of this smoothness difficulty, which physicists usually call the horizon problem. One is to posit that the universe was always smooth. Most physicists find this unacceptable because some sort of gigantic quantum irregularity seems to have been necessary to create the universe out of nothing. Another possibility, inflation theory, argues that the universe went through a brief period of extremely rapid expansion. The explosion of inflation would have proceeded even more rapidly than called for by earlier Big Bang models. Inflation would start with a region of space that was within its own horizon, so that it could have already homogenized. This region was tiny, only about 3×10^{-27} meters across, far smaller than an atomic nucleus. By the end of the inflationary period of accelerated expansion, at roughly 10^{-32} seconds, the universe had grown much larger. The region of space that would later expand to become today's visible universe is still homogenized, which explains why the cosmic microwaves look smooth today.

In this view, however, the entire universe is not necesarily smooth. There may be irregularities beyond today's visible horizon. The inflation theory doesn't guarantee that the universe is smooth; it merely makes it possible for the region of space we can see to have expanded from a region that was smoothed by physical (that is, slower-than-light-speed) processes in the very early stages of the Big Bang. Beyond our view, there could be other regions that underwent inflation differently and are thus cooler or hotter, less dense or more dense than our part of the Universe.

Some of those regions never reached an extent of billions of light-years and so cannot contain life as we understand it, since life requires the evolution of stars to cook the heavy elements.

Another major problem with the unmodified Big Bang that bothers most cosmologists is that some of the most reliable determinations of the universe's mass density give a value one-tenth of "critical" density—the density necessary to make the universe closed and finite. Yet this density is unlikely to be anywhere near critical unless it started out *exactly* at critical density. Both the closed (finite) and the open (infinite) models of the Big Bang require an enormous change in the ratio of the mass density of the universe to critical density between the beginning and now. This ratio is like the ratio of the gravitational energy of the universe to its kinetic energy. In order for this ratio to be near one today, the gravitational and kinetic energy of the earliest Universe would have to have been the same to within about one part in 10^{60}. In other words, the expansion would have to have begun with just the right speed to keep it going forever—but just barely. It is hard to understand how this could have happened accidentally. Perhaps each particle in the universe was and is bound by gravity to the remainder of the Universe with an energy exactly equal to its rest energy—its mass times the speed of light squared.

The difficulty with density is sometimes called the flatness problem because a universe with critical density is flat—that is, it has a curvature of space that is neither positive, as in the closed universe, nor negative, as in the open universe. Inflation theory solves the flatness problem by making the universe so large at such an early time that the observable portion of it is very nearly flat now. (A flat universe is open, but just barely.) It predicts that the average density of matter today is very close to its critical value. Since observed matter is only about one-tenth of critical density, there must be enough dark matter to make up the difference—if, that is, inflation is correct. Inflation theory, as we have seen, is now in doubt because the age of the universe it predicts, using current measurements of Hubble's constant, appears to be less than the age of certain stars. Another problem with the inflation model is that, even in modified form, it predicts irregularities in the early universe too big to agree with the observed smoothness of the cosmic microwaves.

No observations directly confirm the inflation theory. But cosmologists are understandably reluctant to give it up until they have a better theory. Moreover, theorists are fond of grand unified particle theory, the foundation of the inflation theory, since it helps explain the observed excess of matter over antimatter in the universe. Cosmologists are still seeking the perfect theory, but it will almost certainly have to include much of the Big Bang concept as presently understood. As always, the problem in cosmology is getting enough data to put theories to a rigorous test. The Hubble space telescope is producing a flood of images with startling clarity. A whole new generation of huge ground-based telescopes is coming on-line. Improved CCD detectors, adaptive optics, and enhanced computing power are extending astronomers' reach. A week hardly passes without a newspaper article trumpeting a new "farthest galaxy" or "most distant supernova ever seen." With each such discovery, our view of the remote past comes into clearer focus. The search for our origins continues.

Further Reading

General Books on Astronomy

Army, Thomas T. *Explorations, an Introduction to Astronomy.* (Mosby, St. Louis, 1994).

Calder, Nigel. *Violent Universe* (Viking Press, New York, 1969).

Kaufmann, William J. *Discovering the Universe.* (W. H. Freeman and Company, New York, 1993).

Morrison, David and Wolff, Sidney C. *Frontiers of Astronomy* (Saunders College Publishing, Philadelphia, 1990).

Sagan, Carl. *Cosmos* (Ballantine Books, New York 1980).

Schatzman, E. L. *The Structure of the Universe* (McGraw Hill, New York, 1968).

Asteroid and Comet Impacts

Chapman, Clark and Morrison, David. *Cosmic Catastrophes* (Plenum Press, New York, 1989).

Glass, Billy P. *Introduction to Planetary Geology* (Cambridge University Press, Cambridge, 1982).

Hartmann, William K. and Miller, Ron. *The History of Earth* (Workman Publishing, New York, 1991).

Hsu, Kenneth J., *The Great Dying.* (Hartcourt Brace Jovanovich, San Diego, 1986).

Muller, Richard. *Nemesis—The Death Star* (Weidenfeld & Nicolson), New York, 1988).

New Developments Regarding the KT Event and Other Catastrophes in Earth History (Lunar and Planetary Institute, Houston, 1994).

Raup, David M. *The Nemesis Affair, A Story of the Death of Dinosaurs and the Ways of Science* (W. W. Norton, New York, 1986).

Raup, David M. *Extinction, Bad Genes or Bad Luck* (W. W. Norton, New York, 1991).

Sagan, Carl and Druyan, Ann. *Comet* (Random House, New York, 1985).

Taylor, Stuart Ross. *Solar System Evolution* (Cambridge University Press, Cambridge, England, 1994).

Supernova Explosions

Asimov, Isaac. *The Exploding Suns* (Dutton, New York, 1985).

Clayton, Donald C. *Principles of Stellar Evolution and Nucleosynthesis* (McGraw-Hill, New York, 1968).

Fowler, William A. *Nuclear Astrophysics* (American Philosophical Society, Philadelphia, 1965).

Genet, Russell, Hayes, Donald, Hall, Donald and Genet, David. *Supernova 1987A: Astronomy's Explosive Enigma* (Fairborn Press, Mesa Arizona, 1985).

Marschall, Lawrence A. *The Supernova Story* (Plenum Press, New York, 1988).

Murdin, Paul and Murdin, Leslie. *Supernovae* (Cambridge University Press, London, 1985).

Shklovskii, I.S. *Stars, their Birth, Life, and Death* (W. H. Freeman, San Francisco, 1975).

Trimble, Virginia. *Visit to a Small Universe.* (American Institute of Physics New York, 1992).

Trimble, Virginia. *Supernova: Part I and Part II* (Reviews of Modern Physics, 54 and 55, October 1982 and April 1983).

Big Bang Cosmology

Abbott, Edwin A. *Flatland, A Romance of Many Dimensions* (Dover Publications, New York, 1952).

Alfven, Hannes. *Worlds-Antiworlds, Antimatter in Cosmology* (W. H. Freeman, San Francisco, 1966).

Gamow, George. *One Two Three . . . Infinity* (Bantam Books, New York, 1971).

Gardner, Martin. *The Relativity Explosion* (Vintage Books, New York, 1976).

Hawking, Stephen. *A Brief History of Time* (Bantam Books, New York, 1988).

Kolb, Edward and Turner, Michael. *The Early Universe* (Addison-Wesley, Reading, Massachusetts, 1990).

Lemonick, Michael. *The Light at the Edge of the Universe* (Villard Books, New York, 1993).

Lightman, Alan. *Ancient Light, Our Changing View of the Universe* (Harvard University Press, Cambridge, Massachusetts, 1991).

Silk, Joseph. *The Big Bang*, second edition (W. H. Freeman and Company, San Francisco, 1995).

Trefill, James. *Space Time Infinity* (Pantheon Books, New York, 1985).

Thorne, Kip S. *Black Holes & Time Warps.* (W. W. Norton and Company, New York, 1994).

Weinberg, Steven. *The First Three Minutes,* updated edition (Basic Books/ Harper Collins, New York 1988).

Index